SECOND EDITION | REVISED PRINTING
LABORATORY MANUAL FOR

BIOLOGY I

Lalitha Jayant | Owen Meyers | Matthew Geddis | Christine Priano

Kendall Hunt
publishing company

Cover image was taken by Lalitha Jayant in the hills of Munnar, India: The atlas moth, *Attacus atlas* of the family Saturniidae, is native to Southeast Asia and has the largest wingspan of all moths.

Kendall Hunt
publishing company

www.kendallhunt.com
Send all inquiries to:
4050 Westmark Drive
Dubuque, IA 52004-1840

CONTENTS

ACKNOWLEDGMENTS v

LABORATORY EXERCISE 1   MEASUREMENT 1

LABORATORY EXERCISE 2   SCIENTIFIC INVESTIGATION 17

LABORATORY EXERCISE 3   SOLUTIONS 29

LABORATORY EXERCISE 4   ACIDS, BASES, AND BUFFERS 41

LABORATORY EXERCISE 5   MACROMOLECULES I—CARBOHYDRATES 53

LABORATORY EXERCISE 6   MACROMOLECULES II—LIPIDS AND PROTEINS 61

LABORATORY EXERCISE 7   THE MICROSCOPE 73

LABORATORY EXERCISE 8   PROKARYOTES AND EUKARYOTES 85

LABORATORY EXERCISE 9   THE CELL 97

LABORATORY EXERCISE 10   DIFFUSION AND OSMOSIS 101

LABORATORY EXERCISE 11   ENZYMES 113

LABORATORY EXERCISE 12   PHOTOSYNTHESIS 123

LABORATORY EXERCISE 13   CELLULAR RESPIRATION AND FERMENTATION 135

LABORATORY EXERCISE 14   CELLULAR DIVISION 147

LABORATORY EXERCISE 15   DNA ISOLATION AND MODELING 159

LABORATORY EXERCISE 16   TRANSCRIPTION, TRANSLATION, AND PROTEIN
STRUCTURE 171

LABORATORY EXERCISE 17   RESTRICTION ENZYMES 181

LABORATORY EXERCISE 18   DNA FINGERPRINTING 197

LABORATORY EXERCISE 19   DNA CLONING AND BACTERIAL TRANSFORMATION 203

LABORATORY EXERCISE 20   GENETICS 215

APPENDIX 1   LABORATORY RULES AND SAFETY INSTRUCTIONS   **227**

APPENDIX 2   MATERIALS USED IN LABORATORY EXERCISES   **229**

APPENDIX 3   STREET MAP OF LOWER MANHATTAN   **235**

APPENDIX 4   THE LABORATORY REPORT   **237**

APPENDIX 5   THE pH METER   **243**

APPENDIX 6   USE OF THE SPECTROPHOTOMETER   **245**

APPENDIX 7   ELECTROPHORESIS   **247**

APPENDIX 8   SOLUTIONS TO COMPREHENSION CHECKS   **251**

APPENDIX 9   RESTRICTION ENDONUCLEASES   **253**

APPENDIX 10   CONVERSION TABLES   **255**

ACKNOWLEDGMENTS

We would like to thank all BMCC Science faculty for their suggestions and encouragement in preparing this laboratory manual.

Special thanks to Dr. Edith Robbins for her invaluable comments and support.

MEASUREMENT

After completing this exercise, students should be able to do the following:
1. Comprehend the differences between the U.S. and the metric systems of measurement. 2. Convert units of measurement within the metric system. 3. Convert units of temperature between centigrade and Fahrenheit temperature scales. 4. Measure distance, volume, mass, and temperature accurately.

I. INTRODUCTION

Measurement is the process of associating numbers with physical quantities. The process of measuring often requires an instrument that is designed and calibrated for that purpose. Throughout history, different systems of measurement have been used, including the use of body parts to determine length. For example, a person's foot, hand, or arm might have been used to measure distance. Hence, measurement of a given distance would vary greatly depending on the size of a person. As a consequence, it was difficult to compare values of measurement.

Following the Industrial Revolution in the late 18th century, the need for a standardized system of measurement became imperative. Since then, a number of measurement systems have come into existence. In the United States, the familiar system of measurement is the **U.S. system**, derived from older English units of measure (Table 1.1). A more uniformly followed measurement system is the International System of Units, or the **SI system**, which is based on the **metric system** (Table 1.2).

II. THE METRIC SYSTEM

The metric system is the universal system of measurement used by all scientists worldwide. Whereas older units of measure in the English system were based on nature and everyday activities, units of measure in the metric system are derived from scientific principles. For example, the English league was once commonly used to mean the distance that can be walked in one hour. However, in the metric system, the **meter** is defined as the distance that is traveled by light in a vacuum in about $\frac{1}{300,000,000}$ of one second.

A. Converting metric system unit

The metric system is based on the number 10. Each type of measurement is represented by a basic unit (Table 1.1). All larger or smaller quantities are designated by a prefix, each of which represents a multiple or factor of 10 (Table 1.2). For example, the basic metric unit of measure for length is the meter (m). A distance of 1000 meters is a kilometer, which is a distance equal to one meter multiplied by one thousand (kilo). As shown in Table 1.2, a multiple of 10 can also be easily designated as a fraction, as a decimal, or in scientific notation. Having a numerical system based on the number 10 allows simple conversions between very small and very large numbers that are often used in science.

Note that U.S. measurements can also be converted from one unit to another. It is also possible to convert units of measure between the two different systems. Conversion factors within and between the two systems are shown in Appendix 10.

B. Metric measurements

In the laboratory, you will often have to obtain accurate measurements of **length**, **volume**, **mass**, and **temperature**. The basic metric units for these measurements are shown in Table 1.1.

TABLE 1.1 Units of measurement in the US system.

Parameter	Common US Units	Basic Metric Units
Length	inch (in) foot (ft) yard (yd) mile (mi)	meter (m)
Area	square foot (ft²) square miles (mi²) acres	square meter (m²)
Weight	ounce (oz) pound (lb) ton	gram (g)
Volume	teaspoon (tsp) tablespoon (tbsp) cup (c) pint (p) quart (qt) gallon (gal)	liter (L)
Temperature	degree Fahrenheit (°F)	degree centigrade (°C) kelvin (K)*
Time	second (s)	second (s)
Speed	miles/hour (mph)	meter per second (mps)
Energy	calorie (cal)	joule (J)

*0 K (absolute zero) is equal to −273 °C and is sometimes used in scientific experiments.

TABLE 1.2   Unit conversions within the metric system of measurements (Prefixes used in SI system).

Prefix	Multiple	Fraction	Scientific Notation
1 giga (G)	1,000,000,000	–	1×10^9
1 mega (M)	1,000,000	–	1×10^6
1 kilo (k)	1000	–	1×10^3
1 hecto (h)	100	–	1×10^2
1 deca (da)	10	–	1×10^1
Basic Unit (m, L, g)	1	–	1×10^0
1 deci (d)	0.1	1/10	1×10^{-1}
1 centi (c)	0.01	1/100	1×10^{-2}
1 milli (m)	0.001	1/1000	1×10^{-3}
1 micro (μ)	0.000001	1/1,000,000	1×10^{-6}
1 nano (n)	0.000000001	1/1,000,000,000	1×10^{-9}
1 pico (p)	0.000000000001	1/1,000,000,000,000	1×10^{-12}

Length: Length is a measure of distance. The basic metric unit of measure for length is the **meter** (m). In the laboratory, length is measured using a **meter stick** or a **metric ruler**. A meter stick measures one full meter in length and is broken down into 100 equally spaced centimeter increments. Each centimeter is further broken down into 10 equally spaced millimeter increments. The millimeter is the smallest unit of metric measure that can be seen with the naked eye and is equal to 1/1000 of one meter. Meter sticks and metric rulers are often marked with both metric and U.S. measurement scales. Be sure to use the appropriate scale when measuring length.

Length is a one dimensional quantity and represents the linear distance from one point to another. *One side of the cube in Figure 1.1 has a length of 1 cm.*

Area is a quantity that is determined by measurements of length in two dimensions. The area (*A*) of a square can be represented by the formula

$$A = l \times w$$

where *l* is the length on one side of the square and *w* is the width of the square. Area is always represented in square units. *One face of the cube in Figure 1.1 has an area of 1 cm × 1 cm, which equals 1 cm² (one square centimeter, or one centimeter squared).* Note that the units for length and width must be the same before multiplying. If they are not, you must first convert them to the same units before multiplying (see Tables 1.1 and 1.2 and Appendix 10).

Volume: The volume of a solid is a quantity that is determined by measurements of length in three dimensions. The volume (*V*) of a cube can be represented by the formula

$$V = l \times w \times h$$

where *l* is the length on one side of the cube, *w* is the width of the cube, and *h* is the height of the cube. Solid volume is represented in cubic units. *The cube in Figure 1.1 has a volume of 1 cm × 1 cm × 1 cm, which equals 1 cm³ (one cubic centimeter, or one centimeter cubed).*

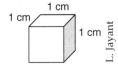

FIGURE 1.1 The cube shown measures 1 cm in length (*l*), 1 cm in width (*w*), and 1 cm in height (*h*).

FIGURE 1.2 Lower meniscus.

Liquid volume is another important measurement in the laboratory. The basic metric unit of measure for liquid volume is the **liter** (L). The volume of a liquid can be measured most accurately using a **graduated cylinder** for larger volumes, and a **pipette** for smaller volumes. A graduated cylinder or pipette contains markings, or graduations, that represent volume measurements. The volume of the contents is determined by viewing the upper surface of the liquid at eye level. A depression, called a **meniscus**, forms on the surface of the liquid (see Figure 1.2). The actual volume measurement is at the point where the lowest part of the meniscus meets the graduated marking. When using a graduated cylinder, this is best seen if the cylinder is set down on a flat table so that the fluid is level.

In the metric system, 1 milliliter (1 mL) of liquid volume is equivalent to the amount of water that will occupy 1 cm³ of space at a temperature of 4 °C.

Note that in the laboratory we use the milliliter, or mL, to measure volume, whereas in medicine, one milliliter of fluid is often designated 1 cubic centimeter, or 1 cc.

Mass: Mass is the amount of matter contained within a substance and can be measured using a **scale**. The basic metric unit of measure for mass is the gram (g). Whereas weight is a property that is influenced by gravity, mass is not. However, because gravity on earth is constant, the mass of a substance on earth is always proportional to its weight. One gram is defined as the mass of one mL of water at a temperature of 4 °C. An important scientific equivalent for water at this temperature is

$$1\text{ g} = 1\text{ mL} = 1\text{ cm}^3$$

Temperature: Temperature is a measure of the hotness or coldness of an object that results from how fast the atoms and molecules in the object are moving. The familiar U.S. unit of measure for temperature is the degree Fahrenheit (°F). In the metric system, the unit of measure for temperature is the degree centigrade, or degree Celsius (°C). Figure 1.3 illustrates the relationship between the Fahrenheit and centigrade temperature scales.

Both Fahrenheit and centigrade scales are based on the temperature at which water freezes and boils. The Fahrenheit scale is divided into 180 equal degree units between the freezing point of water

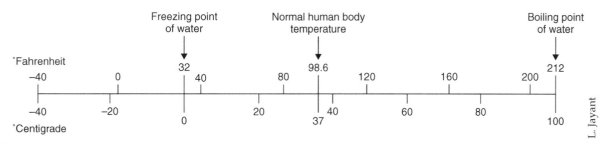

FIGURE 1.3   Comparison of Fahrenheit and centigrade temperature scales.

(32°F) and the boiling point (212°F). The centigrade scale is divided into 100 equal units between the freezing point of water (0 °C) and the boiling point (100°C). A third scale that is often used in science is the kelvin scale, and the unit of measure is the kelvin (K). Like the centigrade scale, the kelvin scale is divided into 100 equal units between the freezing and boiling temperatures of water. However, on the kelvin scale, the freezing point of water is equal to 273.15 K, and the boiling point is 373.15 K.

A temperature of 0 K is equal to –273°C, and is known as **absolute zero**. At this temperature, the motion of all atoms and molecules would hypothetically cease. A temperature of absolute zero has never actually been reached.

The instrument that is used to measure temperature is the **thermometer**. The most common thermometers used in a laboratory are the alcohol thermometer and the mercury thermometer. Either alcohol or mercury is contained within a glass bulb that has a smaller capillary tube extending from it. As temperature increases, the liquid in the bulb expands into the capillary tube. A calibrated scale on the outside of the thermometer spans the length of the capillary tube and correlates the level of the liquid in the tube to a degree measurement for temperature.

Units of temperature measurement can be converted between the different temperature scales. Temperature conversion formulas are shown below. Because one centigrade degree unit is equivalent to 1.8 Fahrenheit degree units, a factor of 1.8 is used when converting temperature measurements between centigrade and Fahrenheit. *Why is the number 32 also part of conversion formulas 1 and 2?* ____

_____

_____

Formula 1.1. To convert from Fahrenheit to centigrade:

$$°C = (°F - 32) \times (100/180)$$
$$= (°F - 32) \times (5/9)$$
$$= (°F - 32) \times 0.556$$

Formula 1.2. To convert from centigrade to Fahrenheit:

$$°F = (°C \times 180/100) + 32$$
$$= (°C \times 1.8) + 32$$
$$= (°C \times 9/5) + 32$$

Formula 1.3. To convert from centigrade to kelvin:

$$K = °C + 273.15$$

C. Practice Problems

1. Convert the following measurements:
   a. 5 km = _____ m
   b. 25 cm = _____ μm
   c. 54 g = _____ kg

   Solution: *a. 5 × 1000 = <u>5000 m</u>*
                 b. 25 × 10 × 1000 = <u>250,000 μm</u>
                 c. 54/1000 = <u>0.054 kg</u>

2. Arrange the following units in order from smallest to largest.
   μm, m, cm, km, Gm, mm, nm, Mm, pm

   Solution: _____

3. Bob runs 2567 m every day. How many km does he run?

   Solution: *1 m = 1/1000 km*
                 2567 m = 2567/1000 = 2.567 km

4. The area of a park measures 843 acres, but your friend is visiting from Canada and wants to know what that area is equal to in square kilometers (km^2). How do you convert the measurement?

   Solution: *1 sq mi = 640 acres; therefore 843 acres = 843/640 = 1.32 mi^2*
                 1 mi^2 = 2.59 km^2; therefore 1.32 mi^2 = 1.32 × 2.59 = <u>3.41 km^2</u>

III. LABORATORY ACTIVITIES

ACTIVITY 1: MEASUREMENT OF LENGTH

In this activity, you will accurately measure length using a meter stick or metric ruler, and convert measurements of length into various units in both the metric and U.S. systems of measurement.

Materials

meter stick
15 cm metric ruler
Appendix 10

Procedure

1. Examine the meter stick. Find the decimeter, centimeter, and millimeter units on the metric scale. *What are the smallest units you can measure using the meter stick?* _____

2. Using the meter stick, measure the exact length of your lab room in meters. Remember, before you can add meters, centimeters, and millimeters, you must first convert all values to meters. Record your measurement in Table 1.3.

3. Refer to Table 1.2 to convert the length of the room into centimeters and then into kilometers. Record these measurements in Table 1.3.

4. Refer to Appendix 10 to convert the length of the room into miles and then into feet. Record these measurements in Table 1.3.

5. Using the 15 cm metric ruler, measure the exact length of your work table. Convert this length into meters, kilometers, centimeters, miles, and feet. Record all measurements in Table 1.3.

Results

TABLE 1.3   Measurements of length.

	length of room	length of table
meters (m)		
kilometers (km)		
centimeters (cm)		
miles (mi)		
feet (ft)		

Which unit do you think should be used to describe the length of the room (cm, m, or km)? Explain.

_____

_____

ACTIVITY 2: MEASUREMENT OF MASS

In this exercise, you will learn to measure mass accurately using a triple beam balance and an electronic balance. You will also convert measurements of mass into various units within the metric and U.S. systems of measurement. Remember that because gravity on earth is constant, the mass of a substance is always proportional to its weight.

A substance is often weighed while inside a container. It is necessary to eliminate the extra weight of the empty container to obtain the actual mass of the specimen. The unwanted weight is called a **tare weight**. In the following activities, you will learn how to balance a scale, how to use a scale to measure mass, and how to eliminate tare weight to determine the accurate mass of a substance.

Materials

triple beam balance   spatula
electronic balance   pebbles
weigh boats   NaCl

Procedure 2a: Measurement of Mass Using the Triple Beam Balance

1. Examine the triple beam balance. On the three beams, find the 1 g, 10 g, and 100 g weights.
2. Balance the scale by first sliding all the weights to the left. Turn the central knob until the arrow points exactly to the zero mark (0 g). The scale is now balanced.
3. Place a pebble on the flat pan of the balance. Notice that the needle moves up.
4. Move the largest weight (100 g) one notch at a time to the right until the needle moves down below the zero mark. Then slide the weight back one notch to the left.
5. Move the next size weight (10 g) one notch at a time to the right until the needle moves down below the zero mark again. Slide this weight back one notch to the left.
6. Slide the last weight (1g) to the right until the needle points exactly to the zero mark.
7. On each beam, read the scale at the point where the weight was placed. Add the gram measures from each beam to find the total mass of the pebble. In Table 1.4, record the mass in grams to the nearest tenth of a gram.
8. Convert the mass of the pebble to centigrams and to milligrams. Record these in Table 1.4.

Results

TABLE 1.4   Mass measurements using the triple beam balance.

units	mass of pebble
grams (g)	
centigrams (cg)	
milligrams (mg)	

Procedure 2b: Measurement of Mass Using the Electronic Balance

1. Examine the electronic balance. Plug in the power cord and turn the power on.
2. Place a weigh boat on the pan and press the "zero" button so that the mass reads 0 g with the weigh boat in place. *This eliminates the tare weight of the weigh boat and balances the scale.*
3. Use a spatula to add NaCl (sodium chloride) until the display on the balance reads 0.500 g.
4. In Table 1.5, record this mass in centigrams and in milligrams.

Results

TABLE 1.5   Weight measurements using the electronic balance.

units	mass of NaCl
grams (g)	
centigrams (cg)	
milligrams (mg)	

ACTIVITY 3: DENSITY

Density is defined as the weight of a substance per unit volume and is measured in g/mL. In this exercise, you will calculate the density of a liquid and of a solid, then compare your results with the rest of the class.

Materials

triple beam balance
10 mL graduated cylinder
50 mL beaker

water
NaCl (table salt)
spatula
weigh boats

Procedure 3a: Find the Density of Water

1. Weigh an empty 50 mL beaker to determine the **tare weight** and record below.
2. Use a graduated cylinder to measure 10 mL of water. Carefully read the volume at the meniscus.
3. Pour the water into the beaker. Weigh the water and beaker together.
4. Calculate the weight of water by subtracting the tare weight of empty beaker from the weight of beaker with water.
5. Calculate the density of water by dividing its weight in grams by its volume in milliliters.

Results

Tare weight (empty beaker) = _____ g
Weight of beaker with water = _____ g
Weight of water = _____ g
Volume of water = _____ mL
Density of water = _____ g/mL

Is your result comparable with that of others in the class? Yes _____ No _____
What is the significance of your result?_____
_____
_____

Procedure 3b: Find the Density of Table Salt (NaCl)

1. Weigh an empty 50 mL beaker to obtain a tare weight.
2. Pour enough salt into a graduated cylinder to reach the 5 mL mark, and transfer the salt into the beaker.

3. Pour the salt into the beaker. Weigh the salt and beaker together.
4. Calculate the weight of the salt by subtracting the tare weight of empty beaker from the weight of beaker with salt.
5. Calculate the density of the salt by dividing its weight in grams by its volume in mL.

Results

Tare weight (empty beaker) = _____ g
Weight of beaker with salt = _____ g
Weight of salt = _____ g
Volume of salt = _____ mL
Density of salt = _____ g/mL
Is this an accurate method for determining density of solids? Explain _____
_____

Is your result comparable with that of others in the class? Yes _____ No _____
Was your density measurement for water or your density measurement for salt more comparable with the measurements of other classmates? What does this tell you? _____
_____
_____

ACTIVITIES TO REINFORCE CONCEPTS
ACTIVITY 4: MEASUREMENT OF TEMPERATURE

In this exercise, you will use a thermometer to measure the temperature of water as it heats. You will also convert temperature from centigrade to Fahrenheit.

Materials

hot plate	glass stirring rod
thermometer	tongs or oven mitts
250 mL beaker	ice
	water

Procedure

> CAUTION: IN THIS EXPERIMENT, THE HOT PLATE AND THE BEAKER WILL GET EXTREMELY HOT. DO NOT USE YOUR HANDS. USE TONGS OR OVEN MITTS TO HANDLE THE BEAKER.

1. Fill a 250 mL beaker with 100 mL ice water and stir with a glass rod.
2. Place a centigrade thermometer into the beaker of water and read the initial temperature. Record this measurement in Table 1.6 below. Remove the thermometer.
3. Carefully place the beaker of ice water on the hot plate.
4. Plug the electrical cord of a hot plate into an outlet and set the heating knob to high.
5. Read the temperature of the water every two minutes for 20 minutes or until the water boils. Be sure to stir the water first with a glass rod before reading the temperature. Remove the thermometer between temperature readings. Record all measurements in Table 1.6.

6. If the water boils before 20 minutes, continue to read the temperature until you get the same reading for 2-4 minutes. Record these measurements.
7. Turn the heating knob on the hot plate to "OFF" and unplug. Let the hot plate and beaker cool before putting away.
8. Using the conversion formulas given above, convert each centigrade temperature reading (°C) to degrees Fahrenheit (°F) and record in Table 1.6.
9. Use Figure 1.4 to graph the rate of temperature change as water heats from its initial temperature to boiling. Add a title. Label *Time (min)* on the x-axis and *Temperature (°C)* on the y-axis.

Results

TABLE 1.6   Temperature Measurements for water.

Time (min)	Temperature (°C)	Temperature (°F)
0		

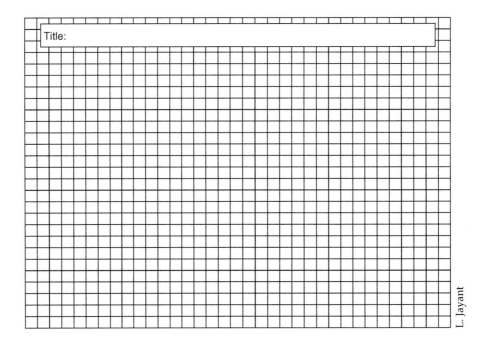

Title:

L. Jayant

FIGURE 1.4

ACTIVITY 5: MEASURING DISTANCE ON A MAP

A **map** is an illustration that depicts the location of different places in relation to one another and in relation to direction (north, south, east, and west). More detailed maps relate the position of a location to the latitude and longitude of the earth. A street map can be viewed as a simple illustration of the ground on paper and is usually drawn looking down from a birds-eye view.

A map has a **key** or **legend** that explains what the symbols on the map designate. The **map scale** relates the measured distance between two points on the map to the actual distance on land. A conversion ratio is given to show how many times the land has been reduced to fit into the map. For example, if the map scale is 1:100,000, then everything on the map has been reduced, or scaled down, one hundred thousand times its original size.

Materials

metric ruler, street map Manhattan (Appendix 3), Appendix 10

Procedure

1. On the street map of Manhattan (Appendix 3), locate the intersection of Chambers Street and Greenwich Street and mark it as BMCC.
2. Locate the Empire State Building at 34th Street and 5th Avenue. Draw a straight line between BMCC and the Empire State Building. Use a metric ruler to measure the length of this line and record below. Use your ruler to measure the scale on the bottom of the map. Determine the conversion ratio for the map and record below.
3. Use the conversion ratio to find the distance in kilometers between BMCC and the Empire State Building. Record below.
4. Refer to Table 1.2 and Appendix 10 to convert this distance into meters, centimeters, and miles. Record below.
5. On the map, draw a walking path between BMCC and the Empire State Building. Use a metric ruler to determine the total length of this route in centimeters. Record below.
6. Using the conversion ratio from step 2, calculate the traveling distance between BMCC and the Empire State Building. Record this distance in kilometers, meters, centimeters and miles.

Results

Distance between BMCC and the Empire State Building as measured on the map:

_____ cm

Conversion ratio for the map: 1 km = _____ cm

Actual linear distance between BMCC and the Empire State Building:

_____ km = _____ m = _____ cm = _____ mi

Walking distance between BMCC and the Empire State Building:

_____ cm = _____ km = _____ m = _____cm = _____ mi

IV. COMPREHENSION CHECK

Metric Measurements

1. 7 centimeters (cm) =_____millimeters (mm)
2. 9 meters (m) =_____centimeters (cm)
3. 8 kilometers (km) =_____meters (m)
4. 5 kilometers =_____meters (m)
5. 5 kiloliters (kL) =_____liters (L)
6. 2 grams (g) =_____milligrams (mg)
7. 40 liters (L) =_____milliliters (mL)
8. 3 kL = _____L
9. 60 kg = _____g
10. 28 m =_____cm
11. 49 m =_____mm
12. 16 L = _____mL
13. 5 kg = _____g
14. 14 cm= _____mm
15. 42 m = _____cm
16. 35 m = _____mm
17. 34 km =_____mm
18. Michele's height is 105 centimeters. Is she taller or shorter than 1 meter? How many centimeters taller or shorter than 1 meter is she?

19. Roberta wants to swim 1 kilometer. How many meters is this?

20. Shiny ran 900 meters. Leona ran 1 kilometer. Who ran further? How much further?

21. Ted lives 400 m from school. Susan lives 3 km from school. How many meters from school does Susan live? Who lives further from school? How much further?

22. Maria is 1.34 m tall. Lynsee is 1300 mm tall. Charles is 141 cm tall. Who is tallest? Who is shortest?

23. One teaspoon holds 5 milliliters. A recipe calls for 2 teaspoons of vanilla extract. How many mL does two teaspoons hold?

24. One bathtub holds about 1 kiloliter of water. Suppose your family uses 10 tubs full of water each week. Approximately how many liters of water would be used in one week?

25. A penny weighs about 3 grams. A dime weighs about 2000 milligrams. Which weighs more? How much more?

26. What is your height in centimeters? In millimeters?

U.S. Distance Measurements (see Appendix 10)

1. 4 ft = _____ in
2. 24 ft = _____ yards
3. 5 yd = _____ ft
4. 1 mi = _____ ft
5. 4 ft 6 in = _____ in
6. 2 yd 2 ft = _____ ft
7. 1 yd 10 in = _____ in
8. 8 ft 4 in = _____ in
9. 3 yd 10 in = _____ in
10. 5 ft 11 in = _____ in
11. 9 ft = _____ in
12. 36 ft = _____ yd
13. 10 yd = _____ in
14. 7 yd 1 ft = _____ ft
15. 4 yd 2 ft = _____ ft
16. Becky can throw a ball 24 yards. Chucky can throw the ball 840 inches. How many feet can each person throw the ball? Who can throw further? How much further?

Common U.S. Weights and Measures (see Appendix 10)

1. 6 cups = _____ pt
2. 8 qt = _____ gal
3. 32 oz = _____ lb
4. 3 qt 1 pt = _____ pt
5. 4 gal 2 qt = _____ qt
6. 6 lb 6 oz = _____ oz
7. 7 qt = _____ pt
8. 18 cups = _____ pt
9. 12 qt = _____ gal
10. 5 gal 2 qt = _____ qt
11. 7 pt 1 cup = _____ cups
12. 5 qt 1 pt = _____ pt
13. Alex bought 6 pints of milk. He wants to give 1 cup of milk to each person in his class. How many people can he serve? How many gallons of milk does he have? How many liters of milk does he have? *(Note: 1 liter is equal to 1.057 quarts).*

14. Milind bought 6 pints of fruit juice. Sara bought 1 gallon of fruit juice. How many quarts of fruit juice did each person buy? Who bought more fruit juice? How many more quarts were there?

SCIENTIFIC INVESTIGATION

After completing this exercise, students should be able to do the following:

1. Describe the steps of the scientific method.
2. Formulate a scientific hypothesis and design a scientific experiment to verify that hypothesis.
3. Identify the controls and the variables of an experiment.
4. Present, analyze, and draw conclusions from experimental results.
5. Write a scientific report.

I. THE SCIENTIFIC METHOD

The process of scientific inquiry, or the **scientific method**, is one of the most fundamental practices in history. Scientific inquiry is based on experiments and observations as opposed to abstract thoughts or ideas. *Sir Isaac Newton's curiosity about an apple falling from a tree is a classic example of an observation that was investigated by the process of scientific inquiry.* The scientific method involves several carefully organized components, which are outlined below:

1. Scientific inquiry often begins with a novel **question** that is raised while observing a natural phenomenon. A scientific question leads to an answer that is both measurable and objective. *"How does air pollution affect the growth of plants?"* is an example of a scientific question. A question that addresses a non-measurable quantity cannot be answered scientifically. *"Is blue a pretty color?"* is an example of a non-scientific question that leads to an opinion. The **purpose** of a scientific experiment is to find the answer to a scientific question.

2. A **hypothesis** is a possible explanation for an observation. The scientist considers all reasonable explanations that might answer a question. A hypothesis can be based on common sense or on prior knowledge about scientific laws and theories. For this reason, a hypothesis is commonly referred to as an **educated guess**. An example of a hypothesis is *"If a plant is exposed to polluted air, its growth will be inhibited."*

3. Each hypothesis, or explanation, is tested by means of a scientific **experiment**. For every plausible hypothesis, there must be a separate experiment. Careful planning of an experiment is a crucial

step in the scientific method. Any factor that can change during the course of an experiment is called a **variable.** The key to designing a successful experiment begins with the identification of all variables relevant to the hypothesis that is being tested. In a **controlled experiment,** only one variable is changed while all others are kept constant. The variables that do not change during the course of an experiment are called **controlled variables**. The variable that is purposely altered or manipulated is called the **independent variable**. Throughout an experiment, the effect of the independent variable is observed. The outcome is the **dependent variable,** which is monitored, measured, or counted in direct response to the effect of the independent variable.

As an example, let us assume that we are studying the effect of air pollution on plant growth. Our question is "How does air pollution affect the growth of plants?" Our hypothesis is "If a plant is exposed to polluted air, its growth will be inhibited." To test this hypothesis, we will proceed to grow plants in the presence and in the absence of sulfur dioxide (SO_2), a common air pollutant. The independent variable is SO_2, which the scientist expects will affect the growth of the plants. To obtain meaningful results, all other factors or variables, such as the amount of water, temperature, and light, must be kept constant at all times. These are the controlled variables that will enable plant growth to be measured in response to only one factor, exposure to SO_2. In this experiment, the dependent variable is plant growth, which can be examined in a number of ways. For example, we can measure the height of the plants, the number and size of the leaves, the number of flowers, or how many seeds are produced. These measurements all define the growth of a plant and can be easily monitored during the course of the experiment.

In summary:

• The independent variable is the factor that is purposely manipulated to cause an effect.
• The dependent variable is the outcome that is observed and can change in direct response to the independent variable.
• The controlled variables are those factors that are kept the same throughout an experiment.

A well-designed experiment should also include a separate **control experiment** that is conducted simultaneously and in which everything is identically duplicated except that the independent variable is omitted. A control experiment ensures that actual experimental results are obtained in direct response to an independent variable, and not in response to external factors.

In the above experiment, plants will be grown over an extended period of time in the presence of SO_2, the independent variable. In a separate control experiment, all conditions will be identical, except that there will be no SO_2 present. The control experiment ensures that any changes observed in the growth of the plants in the test experiment will be a direct result of exposure to SO_2.

After clearly identifying all variables and controls, an **experimental design** is developed in which all **materials** needed for the experiment are described and the **procedure**, also called a **protocol**, is precisely outlined in a step-by-step manner, as in a recipe. The experimental design should also include a means for recording results.

4. The **results** obtained from an experiment are recorded and can be organized and presented in the form of tables, charts, pictures, or diagrams. Results are often analyzed statistically to verify their interpretation. Experimental results are valid only if they can be reproduced. Using the same materials and procedures, one should be able to repeat an experiment several times and obtain similar results.

5. Based on a thorough analysis of the experimental results, a **conclusion** is drawn that states whether the results support the original hypothesis, and future experiments might be discussed. To document the outcome and interpretation of an experiment, a scientific report can be written

that summarizes the experiment and the application of the scientific method (see Appendix 4 for how to write a scientific laboratory report).

The flow chart in Figure 2.1 summarizes the steps of the scientific method. This scheme can be used as a reference when outlining an experiment.

II. PRESENTING, ANALYZING, AND INTERPRETING THE RESULTS OF AN EXPERIMENT

The results obtained during an experiment can be presented in a number of ways. Quantitative or measured results are called **data** (singular, datum). Data are often organized and summarized in tables and then presented in the form of a chart or graph. This visual representation helps to clarify results and aids in interpretation.

Tables

A table is a simple way to collect and organize experimental results (see Figure 2.2). Tables are also used to display results when there are several dependent variables in one experiment. When constructing a table, the following guidelines should be followed:

1. A table should have a title that describes the type of results displayed.
2. All values of the same kind should be placed within a single column.
3. A heading for each column should indicate which variable was measured and the units used for measurement.
4. The data from a control experiment should be included.

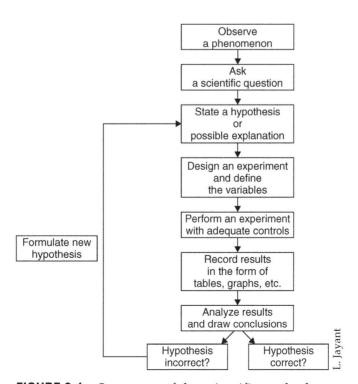

FIGURE 2.1 Summary of the scientific method.

Charts and Graphs

A charts or graph is used to show the direct relationship between the independent variable and the dependent variable. This type of diagram not only provides a visual summary of experimental results, but presents a clear picture of the effect that the independent variable had on the outcome of the experiment (see Figure 2.2). Results can be presented as a line graph, a bar graph, or a pie chart. When constructing a line or bar graph, the following guidelines should be applied:

1. A graph should contain a title that describes what is being shown.
2. The independent variable is placed along on the x-axis (the abscissa, or the horizontal axis).
3. The dependent variable is placed along the y-axis (the ordinate, or the vertical axis).
4. Both axes should intersect at an appropriate point, and the numerical range for each axis should be selected so that the interpretation of the graph is clear. For a linear scale, the increments marked on an axis should be equally spaced.
5. The x-axis and y-axis should be labeled and the units of measurement indicated.
6. A legend should be included to indicate the source of the data.

TITLE   Average height of bean plants.

Air \ Days	Average shoot length (cm) of plants in clean air (control)	Average shoot length (cm) of plants exposed to SO_2 polluted air
0	5.0	5.0
5	6.2	5.6
10	8.8	6.8
15	11.5	6.7
20	17.1	7.3
25	21.7	7.8
30	30.8	8.7

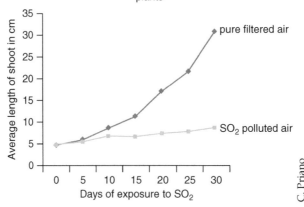

FIGURE 2.2   Example of how to present results in a table (top) and a graph (bottom).

7. The type of graph that is used depends on the type of results to be displayed. If the data are continuous, a line graph should be selected. If the data contain discontinuous measurements or non-numerical categories, a bar graph can be used. Alternatively, percentages can be depicted in a pie chart.

Example of how to present experimental results:

Consider the hypothetical experiment above in which the effect of sulfur dioxide (SO_2) on plant growth was being studied. Figure 2.2 illustrates how data from this experiment might be summarized and presented in a table and a graph.

Summary of the experiment:

1. *Two sets of 50 bean plant seedlings were selected to study the effect of SO_2. Plants were selected that were the same age and height. All plants were grown under the same conditions from seeds of the same parent plant. Each set of plants was placed in a separate glass chamber, one ventilated with air containing a fixed concentration of SO_2 (polluted air), and the other ventilated with clean filtered air.*

2. *The effect of the air on the length of the plant shoot was studied for one month. The type of soil used, the type and the amount of fertilizer used, the amount of light, water and air, and the size of glass chambers were all kept constant.*

3. *What are the independent, dependent, and controlled variables in this experiment? What is the control experiment?*

   <u>*independent variable:*</u> *exposure to SO_2*
   <u>*dependent variable:*</u> *length of the plant shoot*
   <u>*controlled variables:*</u> *bean seedlings, soil, fertilizer, light, water, air, size of chamber*
   <u>*control experiment:*</u> *same experimental setup, but without SO_2*

4. *The shoot length of each plant was measured every five days for one month.*

5. *The results are represented in the table and graph shown in Figure 2.2. Notice how the graph shows a clear and visual representation of the results.*

III. LABORATORY ACTIVITIES
ACTIVITY 1: DESIGN AND CONDUCT A SCIENTIFIC EXPERIMENT

In this exercise, you will follow the steps of the scientific method to do the following:

1. Formulate a hypothesis.
2. Design and conduct an experiment to test your hypothesis.
3. Identify all the variables and controls in your experiment.
4. Present your results in the form of a table and a graph.
5. Discuss the results and draw a conclusion.
6. Write a laboratory report.

Materials

triple beam balance	beakers	water
thermometer	graduated cylinders	salt
weigh boats	stirrer	spatula

Procedure

Work in groups of three or four. Plan an experiment based on the observation given below about ocean water. Use any of the materials provided and what you have learned about scientific measurements. *Note that the salinity of ocean water is 3.5% salt, which is 3.5 grams of salt per 100 mL of water.* (At the instructor's discretion, you may design an alternative experiment based on an observation of your own.)

1. Observation: Objects float better in ocean water than they do in fresh water.
2. What question could you ask about the observation?

   _____

3. Formulate a testable hypothesis that you think will answer your question.

   _____

   _____

4. Design a simple experiment that might support your hypothesis. Identify the variables:

   _____

   _____

   _____

   _____

   _____

   _____

   _____

   Independent variable:_____

   Dependent variable:_____

   Controlled variables:_____

5. Describe your control experiment:

   _____

   _____

   _____

   _____

6. Use Table 2.1 to record your results, and use Figure 2.3 below to graph your results. Be sure to include a title, labels for the x and y axes, and units.
7. Discuss your results with your lab partners and draw a conclusion.

   _____

   _____

   _____

   _____

8. Write a scientific report (see Appendix 4 for how to write a scientific report).

Results

TABLE 2.1 _____

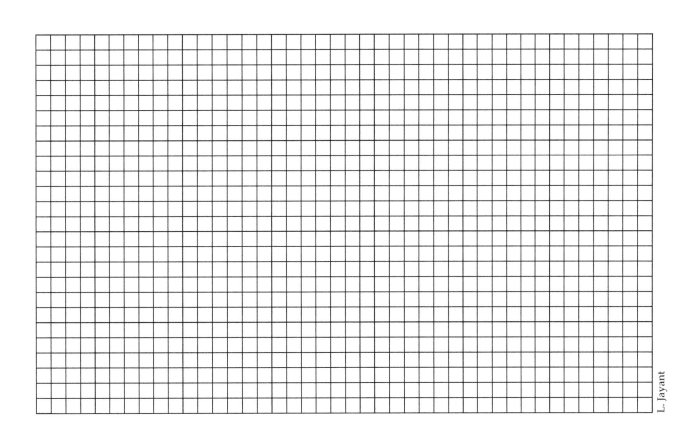

FIGURE 2.3 _____

ACTIVITY TO REINFORCE CONCEPTS
ACTIVITY 2 (OPTIONAL): WRITE A GRANT PROPOSAL

A grant proposal is a document written by the scientist to obtain supportive funds needed to carry out a scientific research project. It presents a plan for conducting one or more experiments that address a specific problem. A well-written grant proposal includes background information about the importance of the topic of interest, as well as the purpose and goals of the research. In outlining the proposal, the scientific method is followed. The question that is being asked is made clear, one or more hypotheses are stated, and an experimental plan is suggested. The design of the experiment must be well thought out and feasible. The possible outcomes or results should be summarized, as well as which of these outcomes will support the hypotheses. If the research proposal is funded, the scientist will carry out the proposed experiments and report the results in a scientific research paper.

Procedure

Work with a partner. Consider the following problem:

Problem: You and your next-door neighbor are determined to get a good harvest of tomatoes in your gardens this season. Together, you both visit the local nursery and buy good quality tomato seedlings. You each plant your seedlings in your respective gardens where there is at least eight hours of full sunlight, and you both water your plants regularly. You are yearning for an abundance of big, ripe, tasty tomatoes. As time goes on, however, you notice that your neighbor's plants are thriving far better than yours. Her plants are healthy-looking and green and have begun to yield a great deal of fruit. Your plants are puny and rather yellow with very few tomatoes. You are very disappointed. You ask your neighbor what she has been doing differently. She tells you that she adds two teaspoons of a commercial fertilizer to her watering can every day before watering her plants. You think that the fertilizer must be the reason that her plants grew so well and yours did not. You want to test your idea.

   Prepare a grant proposal:

1. In an introduction, explain your objectives and provide background information about the importance of your topic. State your question and your hypothesis.
2. Design one or more experiments that will address your hypothesis. Follow the steps of the scientific method. Be sure to identify the dependent, independent, and controlled variables for each experiment.
3. Describe your control experiment.
4. Explain how you will collect data for your results, and how you plan to present them (with tables, graphs, etc.). Explain what kind of results will either support or disprove your hypothesis.

   At the discretion of the instructor, students that present the best proposal will be awarded two plants, soil, and fertilizer. The students will carry out the experiment and present the results in a written report.

IV. COMPREHENSION CHECK

A group of students were studying the effect of a particular plant hormone, giberrelin, on pod production in pea plants. They used 3 cm seedlings that were all grown under identical conditions from seeds of the same parent plant. They grew two sets of 10 plants, one set in the presence and one set in the absence of giberrelin. All plants were grown in a greenhouse at a constant temperature. All were given the same amount of soil, water, and light throughout the course of the experiment. After 12 weeks, the number of pods produced per plant in each group was counted. The results are presented in Table A.

1. Answer the following questions about this experiment:
    a) What is the question that is being asked (the problem)?

    _____

    _____

    b) What is the hypothesis that is being tested?

    _____

    _____

    c) What is the independent variable for this experiment?

    _____

    _____

    d) What is the dependent variable for this experiment?

    _____

    _____

    e) What are the controlled variables for this experiment?

    _____

    _____

TABLE A   Number of pea pods produced after 12 weeks.

PLANTS GROWN WITH GIBERRELIN		PLANTS GROWN WITHOUT GIBERRELIN	
Plant #	Number of Pods	Plant #	Number of Pods
1	19	1	26
2	26	2	25
3	25	3	22
4	17	4	21
5	20	5	21
6	32	6	20
7	24	7	22
8	12	8	19
9	21	9	27
10	23	10	24

_____
_____

f) Describe the control experiment.

_____
_____
_____
_____
_____

2. Use the data in Table A to construct a bar graph showing the average number of pea pods produced by each group of plants. Include a title, labels for the x and y axes, and units.

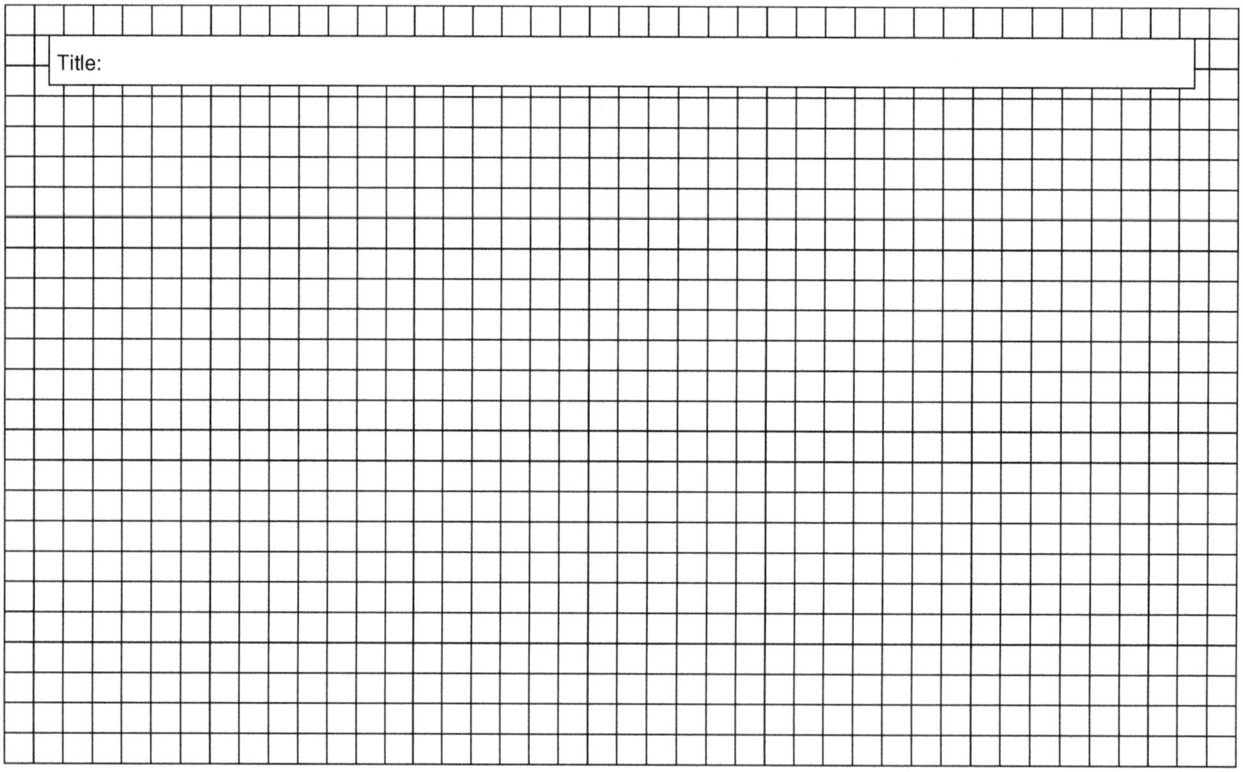

Title: _____

L. Jayant

3. Describe what the results in the graph show.

_____
_____
_____
_____
_____
_____
_____

4. Discuss the possible explanations for the results.

_____

_____

_____

_____

_____

_____

_____

5. Based on their results, what can the students conclude?

_____

_____

_____

_____

_____

_____

_____

LABORATORY EXERCISE 3

SOLUTIONS

After completing this exercise, students should be able to do the following:

1. Identify different types of solutions.
2. Distinguish the difference between a solute and a solvent.
3. Calculate the quantities of solute needed to prepare a solution.
4. Prepare weight-percent and molar solutions.

I. INTRODUCTION

Most experiments in biology use one or more chemical **solutions**. A solution is defined as a mixture of one or more **solutes** dissolved in a **solvent**. A solute is any substance that is dissolved *(for example, sucrose or table sugar)*, and a solvent is the substance in which the solute is mixed *(for example, water)*. Solutions are classified as mixtures because the components retain their chemical properties and can be separated by physical means. Water is an extremely important solvent in nature. In fact, most biological reactions take place in water. For this reason, water is considered a **universal solvent** and is the solvent most often used for biological experiments. Solutions that contain water as the solvent are known as **aqueous solutions**.

Solutions are **homogeneous**, meaning that the distribution of solute is uniform throughout the solution. A solute cannot be separated from a solvent by standing or by gravity. The amount of solute that can completely dissolve in a solvent is characteristic for any given solute/solvent combination. A solution is said to be **saturated** if the solvent has dissolved as much solute as it can possibly hold. A solution is said to be **dilute** if there is less of the solute than the solvent can hold.

Solutions can exist as any state of matter, whether solid, liquid, or gas. Accordingly, there are different terms to describe different types of solutions:

1. **Solid-solid solution**: *for example, 50 micrograms (50 mcg) of levothyroxine is mixed with D-Lactose and a gum to make thyroxine tablets.*
2. **Solid-liquid solution**: *for example, 900 mg of sodium chloride is dissolved in 100 mL of pyrogen-free distilled water to prepare 0.9% normal saline.*

3. **Liquid-liquid solution:** *for example, 5 mL of concentrated acetic acid is dissolved in 95 mL of distilled water to make 5% acetic acid, or table vinegar.*

4. **Gas-liquid solution:** *for example, carbon dioxide (CO_2) is dissolved in pure water to make seltzer.*

5. **Gas-gas solution:** *for example, 5 L of CO_2 is mixed with 95 L of molecular oxygen (O_2) to prepare 100 L of medical oxygen.*

II. PREPARING SOLUTIONS

Many chemical reactions in scientific experiments depend on the types of solutes and solvents that are used as well as the ratio of solute to solvent that is present. Therefore, the success of an experiment often depends on how accurately a chemical solution is prepared. The preparation of a solution involves the following steps:

1. The precise amount of solute must be measured accurately. If the solute is a solid (such as sucrose), its **weight** (w) is measured using a balance or scale. If the solute is a liquid, the **volume** (v) is measured using a graduated cylinder or pipette.

2. The measured solute is mixed with the measured quantity of solvent.

3. In some cases, the pH (acidity or alkalinity) of the solution needs to be adjusted.

The **concentration** of a solution is defined as the amount of solute per unit volume of solvent. More simply put, it is a measure of how much solute is dissolved in a solution. The concentration of a solution can be represented in a number of ways:

1. **Weight-percent** is defined as the weight (w) in grams of a solid solute dissolved in a total volume (v) of 100 mL of solution. *For example, 25 grams of sucrose dissolved in water to a total volume of 100 mL is a 25% (w/v) sucrose solution.*

2. **Volume-percent** (if the solute is a liquid) is defined as the volume of a solute in milliliters (mL) dissolved in a total volume of 100 mL of solution. *For example, 50 mL of 100 % ethanol mixed with 50 mL of pure water is a 50% (v/v) ethanol solution.*

3. **Molarity** refers to the number of **moles** of a chemical substance dissolved in a total volume of 1 liter (1 L) of solvent.

*Note: The **mole** is a fundamental unit of mass used in chemistry. One mole of any substance is an exact measure equal to 6.02×10^{23} molecules of that substance. One mole of a substance is equal to its **molecular** or **atomic mass**, expressed in grams. **Molarity** is a measure of the number of moles of a substance in a solution. In a one molar (1 M) solution, 1 mole of solute is dissolved in 1 liter of solution. Molarity directly relates the number of moles of a solute to the volume of the solution.*

III. LABORATORY ACTIVITIES

In today's laboratory, you will learn how to prepare weight-percent and molar solutions.

ACTIVITY 1: PREPARING WEIGHT/VOLUME (W/V) PERCENT SOLUTIONS

To prepare a (w/v) percent solution, first calculate the required amount of dry solute needed, then dissolve this amount of solute in the appropriate amount of liquid solvent.

Materials

triple- beam balance or electric balance	50 mL graduated cylinder
	beaker
weigh boats	sodium chloride (NaCl)
spatula	water

Procedure 1A: Calculate how to Prepare 50 ml of a 5% (w/v) Solution of Sodium Chloride.

The weight-percent of a solution is the number of grams of solute per 100 milliliters of solution (g/100 mL).

A 5% (w/v) solution of sodium chloride (NaCl) contains 5 g of NaCl in a total of 100 mL of solution, or 5 g/100 mL.

To calculate the number of grams needed to make a given volume of a weight-percent solution, the following formula can be used:

Formula 1:

 # grams of solute needed = [% of solution (g/100 mL) × volume (mL)]

Using this formula, we can now calculate the number of grams of NaCl needed to prepare 50 mL of a 5% (w/v) solution of NaCl:

$$\text{\# grams NaCl} = (5 \text{ g}/100 \text{ mL}) \times 50 \text{ mL} = 2.5 \text{ g}$$

We need 2.5 grams of NaCl to prepare 50 mL of 5% sodium chloride.

Procedure 1b: Prepare 50 mL of a 5% Solution of NaCl.

1. Place a weigh boat on a scale and measure its weight in grams (the tare weight, equal to x).
2. Calculate the tare weight plus 2.5 g (x + 2.5 g) and set the scale at that weight.
3. Slowly add NaCl to the weigh boat using a spatula. When the scale is balanced, you will have weighed out 2.5 g of NaCl.
4. Pour the NaCl into a clean beaker.
5. Rinse the weigh boat with a few drops of water to recover any remaining NaCl and pour into the beaker.
6. Using a 50 mL graduated cylinder, measure about 35 mL of distilled water and pour into the beaker containing the NaCl.
7. Gently swirl the beaker to dissolve all of the NaCl. (Make sure the NaCl is completely dissolved and you cannot see any solid particles.)
8. Pour the NaCl solution back into the 50 mL graduated cylinder.
9. Slowly drip water down the side of the cylinder to bring the total volume up to 50 mL. Be sure to measure the volume of the solution using the lowest part of the meniscus.

Important note: To make this solution, do not simply add 2.5 g of glucose to 50 mL of water. This will introduce error, because adding the solute to the solvent changes the total volume of the solution. As a result, the final percentage of your solution will not be correct.

ACTIVITY 2: PREPARING MOLAR SOLUTIONS (MOLES/LITER)
Materials

triple beam balance or electric balance	100 mL graduated cylinder
	beaker
weigh boats	sodium chloride (NaCl)
spatula	water

The molarity (M) of a solution can be written as:

$$M = \frac{\text{\# moles of solute}}{\text{1 liter of solution}}$$

To prepare a molar solution, we need know the molecular weight of each solute. We can calculate the molecular weight of a solute by adding together the atomic mass of all the atoms that make up a substance. This information can be found in the Periodic Table of Elements, as well as on the container of solute. For example, the molecular weight of NaCl is equal to the atomic mass of Na (22.99 g) plus the atomic mass of Cl (35.45 g), or 58.44 g. This is the actual weight of 1 mole, or one gram-molecular mass, of NaCl. A 1 M (*one molar*) solution of NaCl is equal to 1 mole of NaCl (or 58.44 g) in a total of 1 L of solution.

Examples: The molecular weight of glucose is 180 g, therefore a 1 M solution of glucose contains 180 g of glucose in one liter of solution; the molecular weight of Tris buffer is 121 g, therefore a 1 M solution of Tris buffer contains 121 g of Tris in one liter of solution; the molecular weight of sucrose is 342 g. Therefore, a 1 M solution of sucrose contains 342 g of sucrose in one liter of solution.

Procedure 2a: Calculate how to Prepare 100 mL of a 1 M Solution of NaCl.

From above, we see that to prepare a 1 M solution of sodium chloride, 58.44 g (1 mole) of NaCl must be added to enough water to make a total of 1 L of solution. We can also prepare different volumes of the same molar solution using the following relationship:

Formula 2:

$$\frac{\text{\# grams in solution 1}}{\text{volume solution 1}} = \frac{\text{\# grams in solution 2}}{\text{volume solution 2}}$$

As an example, this formula can be used to calculate the number of grams of NaCl that are needed to prepare 100 mL of a 1 M solution. We already know that a 1 M solution contains 58.44 grams of NaCl in a total volume of 1000 mL. Therefore, using Formula 2, the calculation is as follows:

$$\frac{58.44 \text{ g}}{1000 \text{ mL}} = \frac{\text{\# grams in solution 2}}{100 \text{ mL}}$$

The number of grams of NaCl needed = [(58.44 g × 100 mL)/1000 mL] = 5.844 g.

 ## Procedure 2b: Prepare 100 mL of a 1 M solution of NaCl

1. Place a weigh boat on a scale and measure its weight in grams (this will be the tare weight, equal to x).
2. Calculate the tare weight plus 5.8 g (x + 5.8 g) and set the scale at that weight.

3. Slowly add NaCl to the weigh boat using a spatula. When the scale is balanced, you will have weighed out 5.8 g of NaCl.

4. Pour the NaCl into a clean beaker.

5. Rinse the weighing boat with a few drops of water to recover any remaining NaCl and pour into the beaker.

6. Using a 100 mL graduated cylinder, measure about 70 mL of distilled water and pour into the beaker containing the NaCl.

7. Gently swirl the beaker to dissolve all of the NaCl. (Make sure the NaCl is completely dissolved and you cannot see any particles.)

8. Pour the NaCl solution back into the 100 mL graduated cylinder.

9. Slowly add water down the side of the cylinder to bring the total volume up to 100 mL. Be sure to measure the volume of the solution using the lowest part of the meniscus.

ACTIVITY 3: CALCULATING HOW TO PREPARE SOLUTIONS OF DIFFERENT MOLARITIES

We can prepare solutions that have different molarities using the following relationship:

Formula 3:

$$\frac{\#\ grams/L}{molarity\ 1} = \frac{\#\ grams/L}{molarity\ 2}$$

Problem 1: How many grams of NaCl are needed to prepare 1 L of 0.1 M NaCl?
Using Formula 3, the calculation is as follows:

$$\frac{58.44\ g\ NaCl/L}{1\ M} = \frac{\#\ grams\ NaCl/L}{0.1\ M}$$

Answer: The number of grams of NaCl needed to prepare 1 L of a 0.1 M solution is

$$[(58.44\ g \times 0.1\ M)]/1\ M\ or\ 5.844\ g.$$

Problem 2: Use Formula 3 above to determine how much NaCl is needed to prepare 1 L of the following NaCl solutions:

a) A 0.5 M NaCl solution requires _____ × 58.44 g, or _____ of NaCl.

b) A 2 M NaCl solution requires _____ × 58.44 g, or _____ of NaCl.

c) A 1.7 M NaCl solution requires _____ × 58.44 g, or _____ of NaCl.

Problem 3: Use Formula 3 to calculate the amount of glucose that is needed to prepare 100 mL of a 0.5 M glucose solution. (The molecular weight of glucose is 180 g).

Answer:

Problem 4: Calculate the number of grams of sodium chloride that are needed to prepare 100 mL of 0.5 M NaCl.

First, use Formula 3 to calculate the number of grams needed to make 1 L of a 0.5 M solution:

$$\frac{58.44 \text{ g NaCl/L}}{1 \text{ M}} = \frac{\text{\# grams NaCl/L}}{0.5 \text{ M}}$$

1 L of a 0.5 M solution contains [(0.5 M × 58.44 g)/1 M], or 29.22 g of NaCl.

Next, use Formula 2 to calculate the number of grams needed for 100 mL of solution:

$$\frac{29.22 \text{ g}}{1000 \text{ mL}} = \frac{\text{\# grams in solution 2}}{100 \text{ mL}}$$

Answer: The number of grams of sodium chloride needed to prepare 100 mL of a 0.5 M NaCl solution is

[(29.22 g × 100 mL)/(1000 mL)], or 2.922 g.

ACTIVITY 4: PREPARING DILUTIONS FROM CONCENTRATED STOCK SOLUTIONS

Materials:

500 mL graduated cylinder
water
1 M NaCl

When preparing a solution for a procedure, it is often more efficient to first prepare a concentrated stock solution, then dilute the stock to the final concentration and volume that is needed. The first step in making a dilution properly is to calculate the volume and molar concentration of both the stock solution and the diluted solution. The volume (V) and molarity (M) of the stock solution and the dilution are related as follows:

Formula 4:

$$M_1 \text{ (stock solution)} \times V_1 \text{ (stock solution)} = M_2 \text{ (dilution)} \times V_2 \text{ (dilution)}$$

Procedure 4a: Calculate the Volume of a 1 M NaCl Stock Solution that is Needed to Prepare 500 mL of 0.25 M NaCl.

Using Formula 4, the calculation is as follows:

$$M_1 \text{ (stock)} \times V_1 \text{ (stock)} = M_2 \text{ (dilution)} \times V_2 \text{ (dilution)}$$
$$(1 \text{ M}) \times V_1 \text{ (stock)} = (0.25 \text{ M}) \times (500 \text{ mL})$$
$$V_1 \text{ (stock)} = 125 \text{ mL}$$

To make a total of 500 mL of 0.25 M NaCl, add 125 mL of 1 M NaCl to 375 mL of water.

Procedure 4b: Prepare 500 mL of 0.25 M NaCl from a 1 M Stock Solution.

1. Measure 125 mL of 1 M NaCl into a 500 mL graduated cylinder.
2. Slowly add water down one side of the cylinder to bring the total volume to 500 mL.

Procedure 4c: Calculate the Volume of a 1M NaOH Stock Solution that is Needed to Prepare 75 mL of 0.05 M NaOH.

IV. COMPREHENSION CHECK

1. What is the weight/volume (w/v) percent of a solution that contains 18.0 g NaCl in a total of 90 mL?

2. What is the weight/volume (w/v) percent of a solution containing 5.0 g KCl in a total of 2.0 liters?

3. How many grams of glucose are needed to prepare 250 mL of a 10% (w/v) solution?

4. How many grams of sucrose are needed to prepare 0.125 liters of a 0.15% (w/v) sucrose solution?

5. You need 25 g of glucose. You are given a 5% (w/v) glucose solution. How many mL of the 5% glucose solution do you need to obtain 25 g of glucose?

6. How many mL are needed to obtain 12.0 g of KCl from a 25% (w/v) KCl solution?

7. What is the molarity of a solution that contains 10.0 moles of glucose ($C_6H_{12}O_6$) in a total of 2.0 L?

8. How many moles of KOH are in 725 mL of a 3.2 M solution of KOH?

9. How many grams of NaCl are in 225 mL of a 0.75 M NaCl solution? *(The molecular weight of NaCl is 58.44 g.)*

10. 250 mL of a 5.0 M KCl solution is diluted with water to a final volume of 500 mL. What is the final molarity of the solution?

11. 250 mL of 8% (w/v) NaOH is diluted with 750 mL water. What is the final concentration of the solution? *(The molecular weight of NaOH is 40.0 g/mol.)*

12. 160 mL of water is added to 40 mL of a 1.0 M NaCl solution. What is the final concentration of NaCl?

13. What volume of water must be added to 2 liters of 12% (m/v) KCl to obtain a 4.0% (w/v) KCl solution? What is the total volume of the solution?

14. The composition of TE buffer is 10 mM Tris-HCl, 1 mM EDTA. You are given a stock solution of 1M Tris-HCl and a stock solution of 500 mM EDTA and you are asked to make 100 mL of TE buffer. How many mL of each stock solution will you need?

15. You want to prepare Luria broth (LB)/ampicillin plates. The final ampicillin concentration in the media is to be 100 μg/mL. You first make up 1 L of LB media and 10 mL of a 100 mg/mL ampicillin stock solution. What is the percent of the stock ampicillin solution that you made? How many mL of this stock solution will you add to the LB media to bring the final ampicillin concentration to 100 μg/mL?

16. Calculate the amount of herring sperm DNA required to prepare 50 mL of a 0.4% solution.

17. How much NaCl should you weigh out to prepare 1 L of a 10 M stock solution of NaCl.

18. Using the above stock solution, calculate the volume of stock solution you will need to prepare the following three dilutions:
    a. 100 mL of 1 M NaCl
    b. 20 mL of 5 M NaCl
    c. 500 mL of 0.5 M NaCl

19. You just made up 1 L of Luria broth media. The Luria broth still needs ampicillin to be added to a final concentration of 50 µg/mL. How much of a 5 mg/mL solution of ampicillin should you add to the media?

20. You are beginning work in a new lab as a research student. Your research advisor gives you the task of preparing 500 mL of a solution with the composition given below. How many grams of each of the following compounds will you need to make up this solution?
    a. 100 mM Tris-HCl buffer (mw = 121 g)
    b. 100 mM NaCl (mw = 58 g)
    c. 0.5 % SDS (mw = 288 g)
    d. 0.5 M EDTA (mw = 372 g)

ACIDS, BASES, AND BUFFERS

After completing this exercise, students should be able to do the following:
1. Comprehend the relationship between pH and hydrogen ion concentration.
2. Use various pH indicators to determine the pH of a given solution.
3. Calibrate and use a pH meter to determine the pH of a given solution.
4. Compare pH changes in buffered and unbuffered solutions.

I. INTRODUCTION

A. Defining Acids and Bases

In nature, a very small concentration of water (H_2O) molecules (1×10^{-14} M) will dissociate, or **ionize**, to release an equal number of hydrogen ions (H^+) and hydroxyl ions (OH^-) (1×10^{-7} M each). For this reason, water is considered to be a **neutral** compound.

$$H_2O \;\rightleftharpoons\; H^+ \;+\; OH^-$$

Many solutions are not neutral. An **acid** is a chemical compound that will dissociate in water to release excess hydrogen ions, resulting in a higher concentration of H^+ than OH^-. Alternatively, a **base** is a chemical compound that will dissociate in water to release excess hydroxyl ions, resulting in a higher concentration of OH^- than H^+.

Example: The following chemical reactions show how HCl (hydrochloric acid) dissociates to form the ions H^+ and Cl^-, and how NaOH (sodium hydroxide) dissociates to form the ions Na^+ and OH^-.

$$\text{Acid:} \quad HCl \;\longrightarrow\; H^+ \;+\; Cl^-$$

$$\text{Base:} \quad NaOH \;\longrightarrow\; Na^+ \;+\; OH^-$$

It is often more meaningful to discuss acids and bases according to how they behave in solution with respect to the concentration of H^+ ions. According to the Brönsted-Lowry theory of acids and bases, an acid is considered to be a proton donor, or a substance that can donate a hydrogen ion, or proton, into solution. (In the example above, HCl is an acid because it donates a proton, or H^+, into solution). Consequently, a base is considered to be any substance that can accept a hydrogen ion, or a proton acceptor (In the same example, NaOH is a base because the OH^- that is generated in solution can accept a proton to become H_2O).

Acids and bases have different characteristic properties. For example, acids will corrode metals, are sour to the taste, and will cause litmus dye to turn red. Bases, on the other hand, feel slippery or soapy to the touch, are bitter to the taste, and will cause litmus dye to turn blue.

B. Strength of Acids and Bases

A **strong acid** *(such as HCl)* or a **strong base** *(such as NaOH)* is one that will dissociate completely in water. These dissociation reactions are not reversible.

$$HCl \longrightarrow H^+ + Cl^-$$

$$NaOH \longrightarrow Na^+ + OH^-$$

A **weak acid** *(such as acetic acid, CH_3COOH)* or a **weak base** *(such as ammonium hydroxide, NH_4OH)* is one that will dissociate only partially in water. These reactions are reversible and attain equilibrium.

$$CH_3COOH \rightleftharpoons H^+ + CH_3COO^-$$

$$CH_3COONa \rightleftharpoons Na^+ + CH_3COO^-$$

When an acid gives up a proton (H^+), what remains is called the **conjugate base** of the original acid. *(Cl⁻ is the conjugate base of HCl)*. When a base accepts a proton (H^+), the resulting compound is called the **conjugate acid** of the original base. *(CH₃COOH is the conjugate acid of CH₃COO⁻)*. As a rule, the conjugate base of a strong acid is a weak base, and a conjugate acid of a strong base is a weak acid.

$$HCl \longrightarrow H^+ + Cl^-$$
conjugate acid   conjugate base
(strong)   (weak)

$$CH_3COOH \longrightarrow H^+ + CH_3COO^-$$
conjugate acid   conjugate base
(weak)   (strong)

C. Determining the pH of a Solution

The strength of an acid or base solution is directly related to the hydrogen ion concentration, which ranges from 10^{-1} M (strongest acids) to 10^{-14} M (strongest bases). The pH of a solution is defined as the negative logarithm of the concentration of hydrogen ions, written as:

$$pH = -\log[H^+]$$

Example: If a solution has a hydrogen ion concentration [H⁺] equal to 10^{-3} moles/liter (or 10^{-3} M), then the pH of the solution is determined as follows:

$$pH = -\log[10^{-3}]$$
$$pH = -(-3)$$
$$pH = 3$$

Note that there is an inverse relationship between hydrogen ion concentration and pH. Therefore, as the hydrogen ion concentration increases, the pH of a solution decreases.

Example: Solution A has a hydrogen ion concentration of 10^{-6} M, and solution B has a hydrogen ion concentration of 10^{-8} M. Although solution A has the higher hydrogen ion concentration, solution B has a higher pH (pH = 8) than that of solution A (pH = 6).

In 1909, the Danish biochemist Sören Sörensen introduced what is known as the **pH scale** for measuring acidity. A typical pH scale is shown in Figure 4.1. Values of pH range from 1 to 14, depending upon the hydrogen ion concentration of a solution. A solution with a pH between 0 and 6 is considered to be **acidic** *(containing acid)*, whereas a solution with a pH between 8 and 14 is considered to be **basic** *(containing base)*. A solution that has a pH equal to 7, between the two extremes, is neither acidic nor basic, but is considered to be *neutral. (Pure water is a neutral compound, and has a pH of 7.)*

One way to measure the pH of a solution is to use a **pH meter**. A typical pH meter consists of a glass electrode, a reference electrode, and a voltmeter. The voltmeter measures the potential difference (voltage) between the two electrodes and displays this measurement as a value of pH. See Appendix 5 for details about the pH meter.

Concentration of H⁺ compared to distilled water	pH	Examples of solutions
10,000,000	0	battery acid (hydrofluoric acid [HF])
1,000,000	1	hydrochloric acid (HCl)
100,000	2	lemon juice, vinegar
10,000	3	grapefruit, orange juice, soda
1000	4	acid rain, tomato juice
100	5	tap water, black coffee
10	6	urine
1	7	pure distilled water
0.1 or 1/10	8	sea water
0.01 or 1/100	9	baking soda
0.001 or 1/1000	10	milk of magnesia [Mg(OH)₂]
0.0001 or 1/10,000	11	ammonia solution (NH₄⁺)
0.00001 or 1/100,000	12	soap solution
0.000001 or 1/1,000,000	13	Clorox bleach
0.0000001 or 1/10,000,000	14	liquid drain cleaner (NaOH)

L. Jayant

FIGURE 4.1 The pH scale. Concentration of H⁺ is in moles/liter.

The pH of a solution can also be determined using various types of **pH indicators**. These are compounds that will change the color of the solution depending on its acidity or alkalinity. For example, the organic dye litmus, mentioned above, will appear red in an acid solution, but blue in a basic solution. Similarly, cabbage extract contains pigments that will turn red in an acid and blue in a base. The chemical phenol red, however, will appear red in a basic solution, but yellow in an acid solution.

II. BUFFERS

A **buffer** is a solution that can resist changes in pH when small quantities of acid or base are added to it. Buffers are extremely important for many biochemical processes in which constant pH is essential. *For example, carbonic acid-bicarbonate is a natural buffer that is vital for maintaining acid-base balance in the blood.*

One way to make a buffer is to mix a weak acid with a salt of its strong conjugate base. For example, acetic acid-acetate buffer can be prepared by mixing acetic acid, CH_3COOH *(weak acid)* and sodium acetate, CH_3COONa *(salt of the strong conjugate base)*. Alternatively, a buffer can be made by mixing a weak base with a salt of its strong conjugate acid.

Example: Consider an acetic acid-sodium acetate buffer. The components dissociate as follows:

$$CH_3COOH \rightleftharpoons H^+ + CH_3COO^-$$

$$CH_3COONa \longrightarrow Na^+ + CH_3COO^-$$

The addition of an acid to this buffer will increase the concentration of hydrogen ions, H^+. Excess H^+ will combine with acetate ions (CH_3COO^-) and shift the equilibrium of the first reaction toward the left. This will maintain the H^+ concentration (and thus the pH) in the solution. Similarly, the addition of a base to this buffer will reduce the H^+ concentration. This will shift the equilibrium of the first reaction toward the right, again increasing the hydrogen ion concentration and thus maintaining the pH.

III. LABORATORY ACTIVITIES

ACTIVITY 1: DETERMINING THE pH OF ACIDS AND BASES AT DIFFERENT DILUTIONS

When an acid is diluted, the hydrogen ion (H^+) concentration of the solution decreases. Correspondingly, the pH increases, because pH is inversely proportional to H^+ concentration. Similarly, when a base is diluted, the pH will decrease as the concentration of H^+ increases. In this exercise, we will use pH indicator strips to determine the pH of several diluted samples of both hydrochloric acid (HCl) and sodium hydroxide (NaOH).

CAUTION: CONCENTRATED HCl AND NaOH ARE HIGHLY CORROSIVE AND CAN CAUSE SEVERE BURNS. DO NOT INHALE AND AVOID CONTACT WITH YOUR SKIN OR EYES. ALWAYS HANDLE WITH CARE. INFORM THE INSTRUCTOR OF ANY SPILLS. (REFER TO LAB SAFETY INSTRUCTIONS IN APPENDIX 1).

Materials

1M HCl	test tubes
1M NaOH	test tube rack
pH indicator strips	10 mL graduated cylinder
marking pencil	pipettes

Procedure

1. Find the pH of 1 M HCl by dipping a pH indicator strip into the solution and comparing the color of the strip with color code on the box. Record the pH in Table 4.1 below.
2. Label six test tubes 1 through 6. Fill each with 9 mL of water and place in a rack.
3. To test tube 1, use a pipette to add 1 mL of 1 M HCl. Mix thoroughly with a pipette.
4. For tubes 2 through 6, continue to dilute each successive HCl solution by adding 1 mL of newly diluted solution to the next tube of water. Mix each thoroughly before making the next dilution.
5. For each of the dilutions in tubes 1 through 6, find the pH as described in step 1.
6. Repeat steps 1 through 5, beginning with 1 M NaOH instead of 1 M HCl.
7. After recording all pH values, calculate the hydrogen ion concentration, $[H^+]$, for each sample.

Results

TABLE 4.1   pH as determined using pH indicator paper.

	HCL			NaOH	
Dilution	**pH**	**[H]$^+$**	**Dilution**	**pH**	**[H$^+$]**
1M	0	10^7	1M		
1	1	10^6	1		
2	2	10^5	2		
3	3	10^4	3		
4	5	10^2	4		
5	6	10	5		
6	6	10	6		

ACTIVITY 2: CALIBRATING AND USING A pH METER

A **pH meter** is an instrument that is used to measure the pH of a solution. To obtain an accurate pH reading, it is necessary to first **calibrate** the pH meter. To do this, **standardization buffers** are used that have known pH values. Common standardization buffers that are generally used for calibration are phthalic acid (pH = 4.01), neutral phosphate (pH = 7.01), and borate (pH = 9.18).

Materials

pH meter	distilled water	1M Tris-HCl
Kimwipes	lemon juice	1M HCl
pH 4 standardization buffer	vinegar	1M NaOH
pH 7 standardization buffer	laundry bleach	

Procedure 2a: Determining the pH of Different Solutions Using a pH Meter

1. Turn the pH meter to "ON".
2. Rinse the electrode with distilled water and carefully wipe using a Kimwipe.

3. Immerse the electrode in the pH 7 standardization buffer. Calibrate the electrode at pH 7 as per the manufacturer's specifications for the pH meter. *The pH meter is now calibrated at one point.*

4. Rinse and wipe the electrode as in step 2.

5. Immerse the electrode in the pH 4 standardization buffer. Calibrate the electrode at pH 4 as per the manufacturer's specifications for the pH meter. *The pH meter is now properly calibrated at two points and can be used to measure the correct pH of different solutions.*

6. Rinse and wipe the electrode as in step 2.

7. Immerse the electrode in distilled water. Read the pH and record in Table 4.2. Rinse and wipe the electrode as in step 2.

8. Repeat step 7 for lemon juice, vinegar, laundry bleach, and baking soda. Record your results in Table 4.2. *Be sure to rinse and wipe the electrode after each reading.*

9. When you are finished measuring the pH of each sample, return the electrode to the storage buffer. *This step is important: if the electrode is not immersed, it will dry out. This will cause permanent damage.*

10. Calculate the hydrogen ion concentration, $[H^+]$, for each sample and record in Table 4.2.

Results

TABLE 4.2 pH of materials as determined using a pH meter.

Material	pH	$[H^+]$

Procedure 2-b: Adjusting the pH of 1M Tris-HCl to pH = 6

In biology and chemistry laboratories, it is essential that solutions used for chemical reactions have the correct pH. *For example, in a protein solution that is not at its optimum pH, the protein might become denatured.* It therefore becomes important to know how to properly adjust the pH of any given solution. pH is adjusted by slowly adding either an acid or a base and monitoring the change in pH using a pH meter.

1. Rinse the pH meter electrode with distilled water. Carefully wipe the electrode using a Kimwipe.

2. Use the pH meter to read the pH of 1M Tris-HCl. Determine whether the pH needs to be decreased or increased to obtain a pH of 6.

3. Rinse and wipe the electrode as in step 1.

4. Add one drop of either 1M HCl (if the pH needs to be decreased) or 1M NaOH (if the pH needs to be increased), swirl gently, and read the pH. Rinse and wipe the electrode as in step 1.

5. Repeat step 4 until the pH equals 6. *Be sure to rinse and wipe the electrode after each reading.*

✶ ACTIVITY 3: DETERMINING pH USING RED CABBAGE EXTRACT, A pH INDICATOR

Indicators are solutions that change color in acid or base solutions. Red cabbage extract is one such indicator. It will not only show if a solution is acidic or basic, but it will indicate the strength of the acid or base. At different pH values, red cabbage extract will turn different colors. Table 4.3 shows an array of color changes observed when red cabbage extract is added to an acid or a base. This chart can be used to determine the approximate pH of a solution.

Materials

pH meter · distilled water
blender · red cabbage
sieve · lemon juice
beaker · vinegar
test tubes · laundry bleach
test tube rack · dissolved baking soda
graduated cylinders · balance
pipettes

Procedure

1. Prepare an extract of red cabbage by blending 10 g of red cabbage and 100 mL of water in a food blender.

2. Strain the mixture through a sieve and collect the extract in a beaker.

3. Label five test tubes 1 through 5 and place in a test tube rack.

4. Add 12.5 mL of the indicated sample to each tube: distilled water (tube 1); lemon juice (tube 2); vinegar (tube 3); laundry bleach (tube 4); dissolved baking soda (tube 5).

5. To the sample in each test tube, add 0.5 mL of red cabbage extract and mix thoroughly using a pipette.

6. Refer to the color chart in Table 4.3 to determine the approximate pH of each sample. Record your results in Table 4.4 below. For bleach, be sure to observe the color change immediately after adding the extract.

7. Determine the pH of each sample using the pH meter. Record the pH in Table 4.4.

8. Observe how the pH of each sample compares when determined using either cabbage extract or the pH meter.

TABLE 4.3   Color changes in red cabbage extract at different pH values.

color of extract:	red	fuschia	purple	blue	blue green	green
approximate pH:	2	4	6	8	10	12

Results

TABLE 4.4 pH values determined using cabbage extract versus a pH meter.

Material	Extract Color	pH Using Cabbage Extract	pH Using the pH Meter
distilled water	purple	purple 6	
lemon juice		fushia 4	
vinegar	~~pink~~ fushia	fushia 4	
laundry bleach		Nochy	
baking soda	blue	Blue 8	

DISCUSSION

Discuss with your lab partners how the pH values determined using cabbage extract compare with the pH measured with the pH meter. Explain in the space provided below.

_____

_____

_____

_____

_____

ACTIVITY 4: EFFECT OF ACID OR BASE ON THE pH OF BUFFERED VS. NON-BUFFERED SOLUTIONS

Most biochemical reactions require a neutral pH. Phosphate buffers present in living organisms help to keep biological solutions close to a pH of 7.

Materials

pH indicator strips   distilled water
beakers   1M phosphate buffer
graduated cylinders   0.1 N HCl
pipettes   0.1 N NaOH

Procedure

1. Fill one beaker with 50 mL of plain water and fill another with 50 mL of 1 M phosphate buffer.
2. Using the pH indicator strips, determine the pH of pure water and of 1 M phosphate buffer. Record your results in Table 4.5 below.
3. To each sample, add a few drops of 0.1 M HCl and mix thoroughly using a pipette.
4. Repeat step 2 to determine the pH of each sample. Record the pH in Table 4.5.
5. To each sample, add a few drops of dilute NaOH and mix thoroughly using a pipette.
6. Repeat step 2 to determine the pH of each sample. Record the pH in Table 4.5.

Results

TABLE 4.5   pH of sample solutions before and after addition of acid or base.

	pH of Water Sample	✹ pH of Phosphate Buffer Sample
Before addition of HCl		
After addition of HCl		
Before addition of NaOH		
After addition of NaOH		

What happened to the pH of water and the pH of phosphate buffer after either acid or base was added?

_____

_____

_____

_____

_____

_____

IV. COMPREHENSION CHECK

1. Describe the properties of an acid.

2. Describe the properties of a base.

3. Explain how the concentration of hydrogen ions in a solution is related to the acidity of the solution.

4. What is pH?

5. What is a buffer?

6. How does a buffer work?

7. Describe how a pH indicator can be used to measure pH.

8. If the pH of a solution is 6, what is the concentration of hydrogen ions in moles/L?

9. What is the pH of a solution that has a hydrogen ion concentration of 1×10^{-8} moles/L?

10. A solution of hydrochloric acid (HCl) has a pH of 2. How much would you have to dilute this solution to obtain a solution that has a pH of 5?

MACROMOLECULES I—
CARBOHYDRATES

After completing this laboratory exercise, students should be able to do the following:

1. Recognize the characteristics of the carbohydrates and be able to use specific tests to identify each sub-group: monosaccharides, disaccharides, and polysaccharides.
2. Identify the chemical reactions that each carbohydrate group can undergo.
3. Test for specific groups of carbohydrates.
4. Analyze an unknown solution or compound and determine the carbohydrate group to which it belongs based on the chemical properties of its components.

I. INTRODUCTION

There are four major types of macromolecules found in living organisms: **proteins, carbohydrates, lipids,** and **nucleic acids**. A macromolecule is a large molecule composed of many smaller subunits called **monomers**. Monomers are linked together in a process called **dehydration synthesis**, where the equivalent of one molecule of water is removed from between two monomer subunits to form a covalent bond. The covalent bond that is formed between the monomers often has a special name.

Chemically reactive parts of large molecules are specific structures called **functional groups**. Each type of biological molecule has characteristic functional groups. You will study each of the four types of biological macromolecules and the chemical reactions that each can undergo. Within each group, there are subtypes of molecules that you will learn to identify. In this laboratory exercise, you will concentrate on the carbohydrates and their properties.

A. Carbohydrates

Carbohydrates are more commonly known as starches and sugars. You will study three groups of carbohydrates: **monosaccharides, disaccharides,** and **polysaccharides**. All carbohydrates have the general formula $(C_nH_{2n}O_n)$. There are several functional groups that are associated with carbohydrates. Among these are the **hydroxyl group (–OH)**, the **ketone group (–CO)**, and the **aldehyde**

53

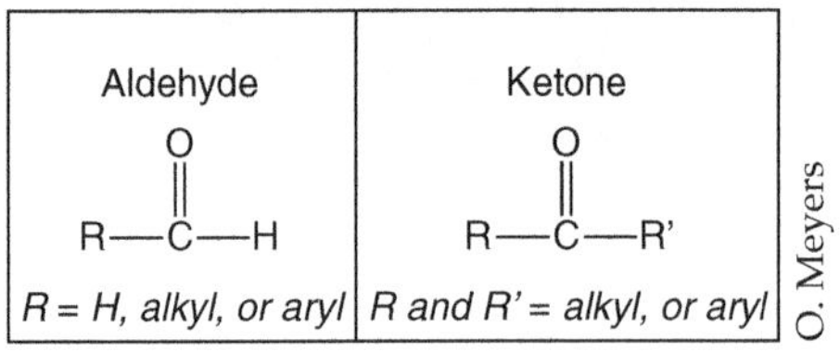

FIGURE 5.1 Structure of an aldehyde and a ketone.

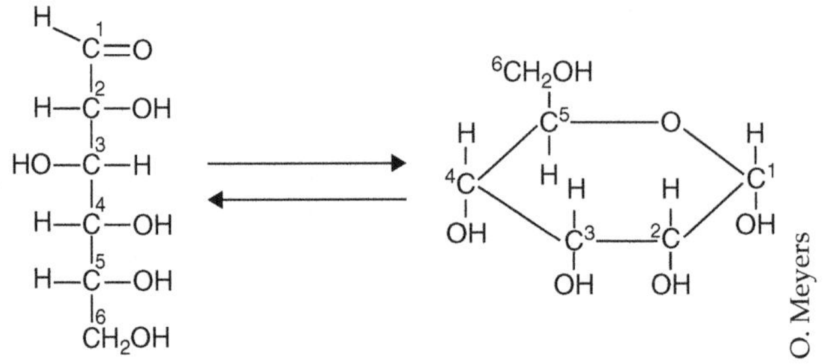

FIGURE 5.2 The transition between the straight chain and the ring structure of a monosaccharide.

group (–CHO) (Figure 5.1). In a structural formula, a functional group is drawn with a line (–) to indicate where a covalent bond attaches the functional group to the rest of the molecule. The remainder of the molecule is designated with the symbol –R.

B. Monosaccharides

The monosaccharides are the smallest carbohydrate molecules. These are the monomer subunits, or building block molecules, that are linked together to form larger carbohydrate polymers. They are often referred to as simple sugars. Specific examples of monosaccharides are **glucose, fructose,** and **galactose**. Many monosaccharides can exist as either ring structures or straight chains and can easily transition between these two structures (Figure 5.2). Most monosaccharides, however, exist almost entirely in the ring form when in solution. Common monosaccharides are six carbon sugars (such as glucose, fructose, and galactose), but there are three, four, and five carbon sugars as well. The two most important of the five carbon sugars are deoxyribose, found in DNA, and ribose, found in RNA.

C. Disaccharides

Disaccharides are formed when two monosaccharides join by dehydration synthesis (loss of water). The covalent bond that is formed is called a **glycosidic bond**. Glycosidic bonds can be either alpha (α) or beta (β), depending on the orientation of the hydroxyl group at the number 1 carbon (Figure 5.3).

Figure 5.4 shows a molecule of the disaccharide sucrose, more commonly known as table sugar. A molecule of sucrose is composed of one molecule of glucose joined to one molecule of fructose by an alpha glycosidic bond. Other common disaccharides are maltose (composed of two molecules of glucose) and lactose (composed of one molecule of glucose and one molecule of galactose).

D. Polysaccharides

Polysaccharides are formed when 3 or more carbohydrates are joined in a long chain by dehydration synthesis (Figure 5.5). Polysaccharides provide both energy and structure for organisms. **Starch**, a common plant polysaccharide, is used for energy storage. Starch is composed of branched chains of glucose molecules connected by alpha glycosidic linkages (Figure 5.5a). In humans, the polysaccharide

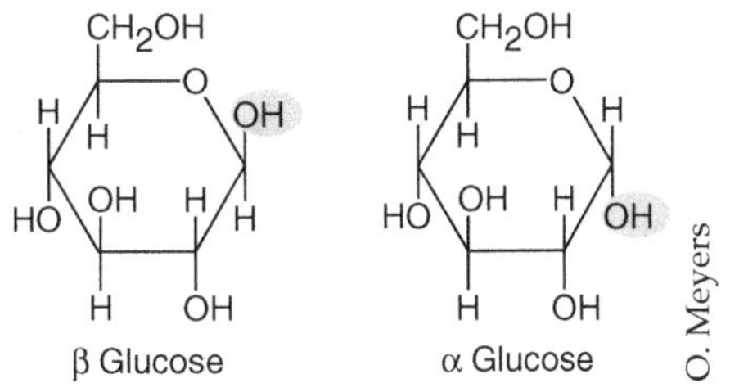

FIGURE 5.3   Example of beta-glucose (the –OH group of carbon 1 points up) and alpha-glucose (the –OH group of carbon 1 points down).

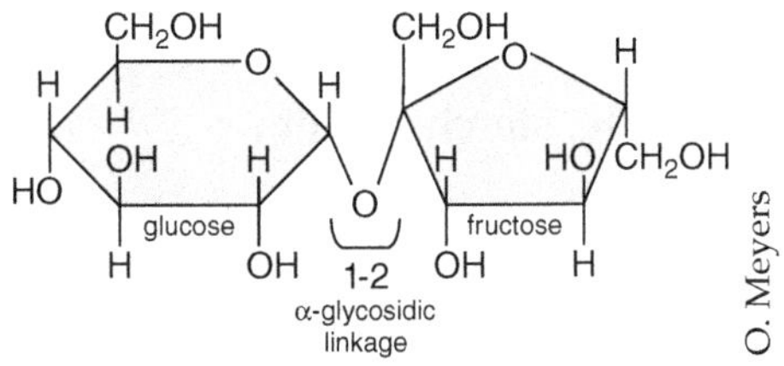

FIGURE 5.4   A sucrose molecule composed of linked glucose and fructose.

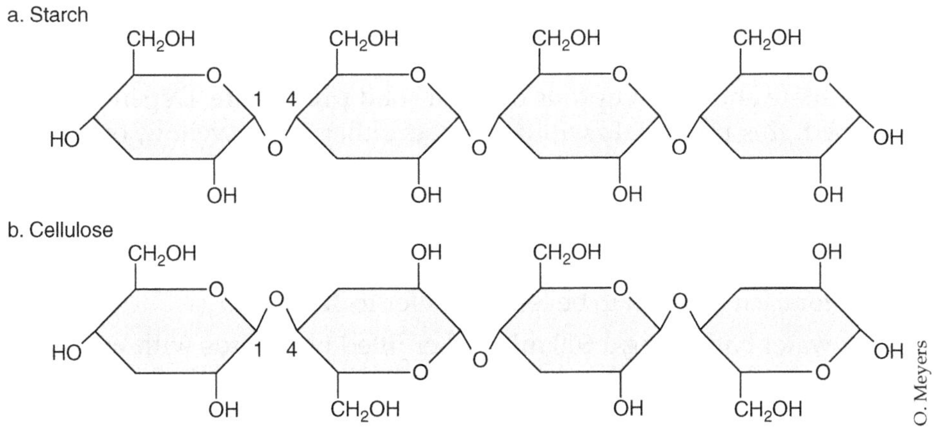

FIGURE 5.5   Starch (a) contains α-glycosidic bonds and cellulose (b) contains alternating α and β-glycosidic bonds. Note the positions of the –CH₂OH and –OH groups.

glycogen stores energy in both liver and muscle cells. Like starch, glycogen is also composed of branched chains of glucose molecules, but it is far more branched than starch. **Cellulose** is an important structural polysaccharide found in plants. In the absence of a skeleton, cellulose serves to give plants their rigid shape. The beta glycosidic linkages in cellulose render this macromolecule indigestible in humans (Figure 5.5b).

II. LABORATORY ACTIVITIES

ACTIVITY 1: BIOCHEMICAL TESTING FOR CARBOHYDRATES

In this exercise, you will perform three common tests that distinguish different types of carbohydrates. Using these tests, you will identify the types of carbohydrates present in a number of common substances, and which carbohydrates, if any, are present in an **unknown** solution.

Materials

Reagents:
Benedict's reagent
Barfoed's reagent
Lugol's iodine

test tubes
hot plates
tongs
test tube holders
test tube rack

Known Solutions:
1% sucrose, glucose, fructose, lactose, starch, water

marking pencil
stirring rods
spot dish
wax pencil or marker

Common Substances:
potato juice, onion juice, Karo syrup, white bread, apple

disposable pipettes
10 mL graduated cylinder
600 mL beaker

unknown carbohydrate solution
unknown carbohydrate solids

Procedure 1a: Benedict's Test for Reducing Sugars

A reducing sugar is one with a free or potentially free aldehyde or ketone group. Benedict's reagent contains copper sulfate, which will react with a reducing sugar at a pH of 8 or higher. The sugar will reduce the copper sulfate and form cuprous oxide, a solid **precipitate**. Depending on the amount of cuprous oxide formed, this precipitate will be seen as either green, yellow, orange, reddish brown, or red. The color change corresponds to the relative concentration of reducing sugar present in the solution. A green color indicates a relatively low concentration of sugar, and a red color indicates a relatively high concentration.

1. Label test tubes for each solution to be tested. (Refer to Table 5.1.)
2. Set up a boiling water bath using a 600 mL beaker filled two-thirds with water.
3. Into each labeled test tube, add 1 mL of the solution to be tested.
4. To each test tube, add 5 mL of Benedict's reagent.
5. Place five test tubes at a time in the boiling water bath for 5 minutes. *Do not attempt to place more than five test tubes into the water bath at the same time.*
6. After 5 minutes, remove the test tubes and allow them to cool to room temperature.
7. Notice whether there is a precipitate and whether there is any color change. Record your observations in Table 5.1.

Procedure 1b: Barfoed's Test

Barfoed's reagent can be used to distinguish monosaccharides from both disaccharides and polysaccharides. Barfoed's reagent is similar to Benedict's reagent in that both contain copper ions, but Barfoed's reagent tends to work at a lower pH, and the reaction requires less time. Reactions should not be heated for more than 2 minutes, as this might cause hydrolysis of disaccharides into monosaccharides and give a false positive result for reducing monosaccharides.

1. Label test tubes for each of the solutions to be tested. (Refer to Table 5.1.)
2. Set up a boiling water bath using a 600 mL beaker filled two-thirds with water.
3. Into each labeled test tube, add 1 mL of the solution to be tested.
4. To each test tube, add 3 mL of Barfoed's reagent.
5. Place five test tubes at a time in the boiling water bath for only 2 minutes. *Do not attempt to place more than five test tubes into the water bath at the same time.*
6. After 2 minutes, remove the test tubes and allow them to cool to room temperature.
7. Notice whether there is a precipitate and whether there is any color change. Record your observations in Table 5.1.

✩ Procedure 1c: Iodine Test

This test is performed to distinguish polysaccharides from monosaccharides and disaccharides. Only polysaccharides will react with iodine to turn blue-black in color.

1. Label the wells of a spot dish (depression dish).
2. To one well in the spot dish, add 0.5 mL (2–5 drops) of starch solution.
3. Add 2 or 3 drops of iodine solution to the starch and notice the color.
4. To each of the other wells, add 0.5 mL (2–5 drops) of the remaining solutions to be tested. (Refer to Table 5.1.) Use a separate well for each different solution.
5. To each well, add 2 or 3 drops of iodine.
6. For each solution, indicate if there was a color change and whether this was a positive or negative result for the presence of starch. Record your observations in Table 5.1.
7. For Procedures 1a, 1b, and 1c, observe the results you obtained for the known solutions (shaded rows in Table 5.1). Using this information, identify the types of carbohydrates present in each of the common substances, and in your unknown solution.

Results

TABLE 5.1   Results of carbohydrate tests.   ✗

Samples	Benedict Test Color Observed	Barfoed's Test Precipitate? (yes or no)	Iodine Test Blue/Black? (yes or no)
Water	Blue		No
Glucose	tomato red		No
Fructose	Brick red		No
Maltose	Brick red		No
Lactose	Brick red		No
→ Sucrose	Blue		No
Starch	Blue polysaccharid		
Karo syrup	Brown		
Potato juice	Blue		
Onion juice	Brown		
White bread			
Apple			
Unknown solution			

Identify the carbohydrate(s) present in your common substances: _____
_____
_____
_____

Identify the carbohydrate(s) present in your unknown solution: _____

Explain how you determined the identity: _____
_____
_____

ACTIVITY TO REINFORCE CONCEPTS

ACTIVITY 2: DETERMINATION OF AN UNKNOWN SUBSTANCE

Students will work in groups of two or three. Each group will be given a solid unknown and determine what type of carbohydrate it contains.

Students will construct a flow chart to plan their experiments (refer to Laboratory Exercise 2: Scientific Investigation). Students will determine which tests to use and outline their procedures. This will be written out by *each* student and is to be either handed to the instructor at the end of the lab session or included with your laboratory report.

Identify your solid unknown: _____

III. COMPREHENSION CHECK

1. Describe the differences and the similarities between monosaccharides, disaccharides, and polysaccharides.

2. What is the monomer subunit for carbohydrates? What types of macromolecules can form by linking carbohydrate monomer subunits?

3. Why are carbohydrates important in living organisms?

4. Why does glucose yield a positive Benedict's test, but starch does not?

5. Name three different monosaccharides and three different disaccharides. What monomer subunits make up the disaccharides that you listed?

6. Name three types of polysaccharides found in living organisms. Describe their functions and their monomer subunits.

7. You are conducting a qualitative analysis of an unknown solution. A Benedict's test results in a color change from blue to green. A Barfoed's test shows no change in color. Lugol's iodine turns a sample of the solution blue/black. Based on these results, what can you conclude about the contents of the unknown solution?

8. A solution gives a positive Benedict's test and a negative Barfoed's test. How could this happen?

9. In the Benedict's and the Barfoed's tests, why is the boiling time critical?

MACROMOLECULES II—LIPIDS AND PROTEINS

After completing this exercise, students should be able to do the following:
1. Recognize the properties of lipids and proteins and be able to use specific tests to identify each. 2. Identify the chemical reactions that lipids and proteins undergo. 3. Test for lipids and proteins in various sample substances. 4. Analyze an unknown solution and determine its composition. 5. Determine the concentration of proteins in a solution.

I. INTRODUCTION

In this laboratory exercise, you will study lipids and proteins, as well as some chemical reactions they undergo. Within each group, there are subtypes of molecules that you will learn to identify.

A. Lipids

Lipids are a group of biological molecules that include the **triglycerides** (fats, oils, and waxes), **phospholipids**, and **steroids**. A common property of all lipids is that they are either fully or partially non-polar, making them **hydrophobic** or insoluble in water. Triglycerides are important for energy storage and are composed of one molecule of glycerol covalently bound to three molecules of fatty acid (Figure 6.1). Phospholipids are important components of cell membranes and differ from triglycerides in that they have a phosphate group (PO_4) that replaces one of the fatty acid chains (Figure 6.2). The phosphate group gives a phospholipid a polar head that is hydrophilic or water soluble, while the fatty acid tail remains hydrophobic. Steroids are lipids that contain four cyclic hydrocarbon rings (Figure 6.3). One of the most important steroids in our bodies is cholesterol, the basis for many steroid hormones. Some lipids can exist as either solids (fats and waxes) or liquids (oils) at room temperature.

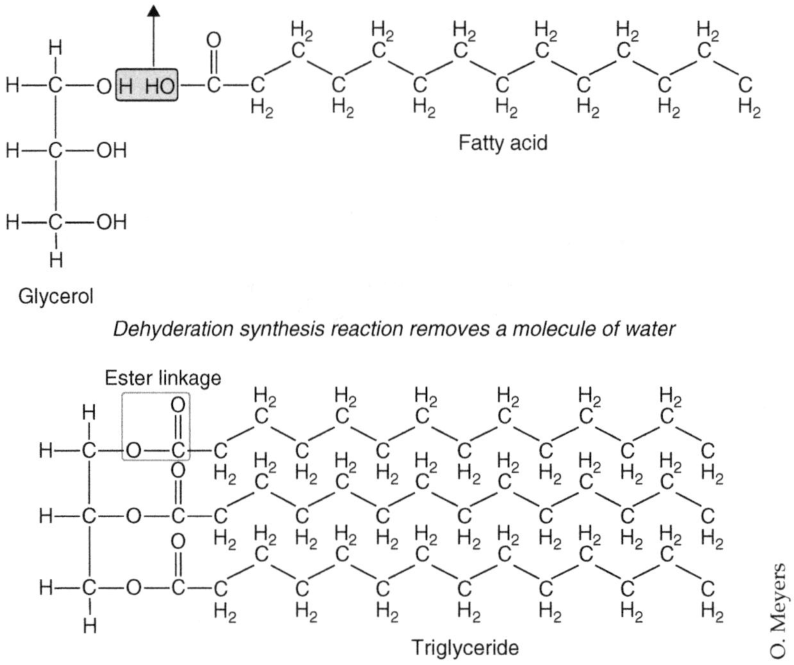

FIGURE 6.1 Structure of a triglyceride.

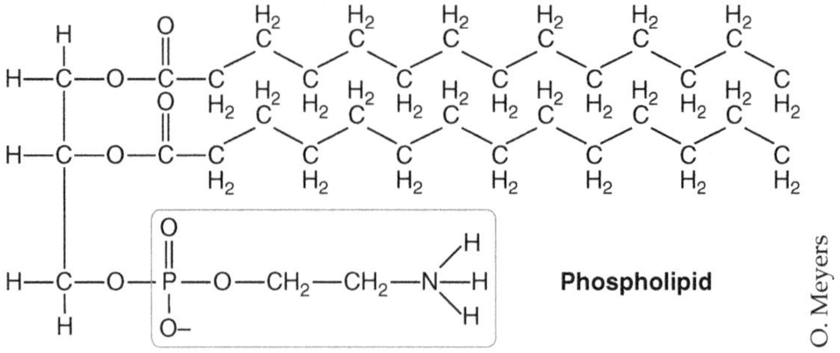

FIGURE 6.2 A phospholipid. Note that it is similar to a triglyceride, but there is a phosphate group in place of one fatty acid molecule.

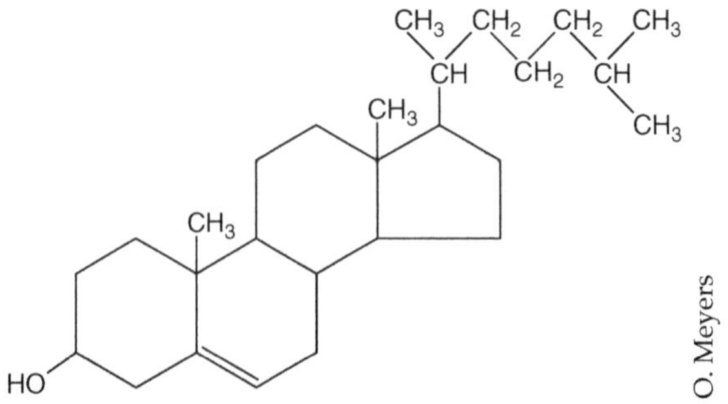

FIGURE 6.3 Cholesterol, a typical steroid.

B. Proteins

Proteins are formed by the joining of **amino acids** by dehydration synthesis. All proteins contain a number of amino acids in a specific sequence. There are a total of 20 amino acids, and every amino acid contains the following functional groups: an amino group (NH_2), a carboxyl group (COOH), and

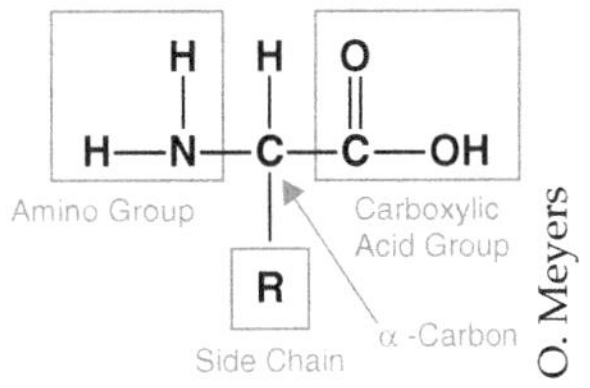

FIGURE 6.4   Structure of an amino acid.

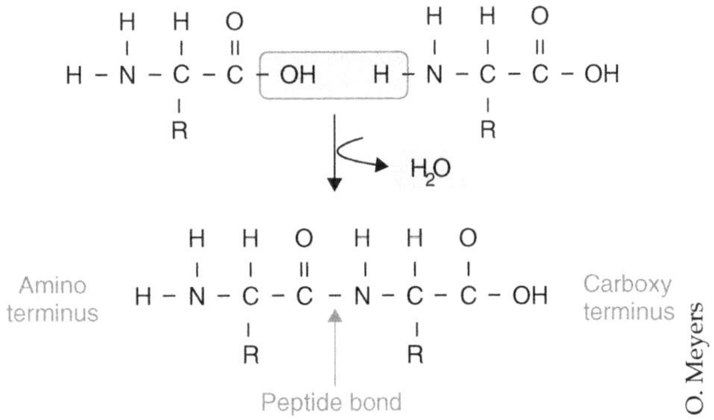

FIGURE 6.5   A peptide bond forms when two amino acids join together to form a dipeptide.

an alkyl group (R) that differs from one amino acid to another (Figure 6.4). The covalent bond formed between successive amino acids is called a **peptide bond** (Figure 6.5). A chain of 50 or more amino acids is called a **protein**. Chains of fewer than 50 amino acids are referred to as **polypeptides**. Proteins have many functions. Some examples of proteins are hormones, transport molecules, receptors on the cell membrane, antibodies, and enzyme catalysts for chemical reactions in the body. Proteins can also be structural in nature. Hair, feathers, nails, wool and silk are common examples of structural proteins.

II. LABORATORY ACTIVITIES

ACTIVITY 1: TESTING FOR LIPIDS

Materials

Reagents:
Sudan IV powder

Test Solutions:
water, oil

Common substance:
white bread

unknown lipid solution 1

Supplies:
disposable test tubes
test tube racks
wax pencil or marker
brown paper towel
1 mL pipettes
5 mL pipettes
pipette dispenser

⚥ Procedure 1a: Sudan IV Test

Sudan IV is a powdered red dye that dissolves only in hydrophobic substances. Sudan IV will stain skin and clothing. Only a small amount should be used and it should be handled with care.

1. Label three test tubes 1, 2 and 3.
2. To test tube 1 add 1 mL of oil and *a few* grains of Sudan IV.

3. To test tube 2 add 1 mL of water and *a few* grains of Sudan IV.
4. To test tube 3 add 1 mL of water, 1 mL of oil and *a few* grains of Sudan IV.
5. Shake each test tube and let sit for one minute.
6. Observe and record your results for test tubes 1-3 in Table 6.1.
7. Label two test tubes numbers 4 and 5, and add 1 mL of the white bread mixture to each.
8. Label two test tubes numbers 6 and 7, and add 1mL of the unknown lipid solution to each.
9. To tubes 4 (white bread) and 6 (unknown), add 1 mL of water.
10. To tubes 5 (white bread) and 7 (unknown), add 1 mL of oil.
11. To each of the four tubes (numbers 4, 5, 6, and 7), add *a few* grains of Sudan IV. Observe the contents in each tube and record your results in Table 6.1.

Results

TABLE 6.1 Results of the Sudan IV Test.

Test Tube	Samples	Observations
1	oil only	C
2	water only	
3	oil plus water	
4	white bread plus water	Cloudy
5	white bread plus oil	Thick parridy at the top
6	unknown plus water	Partidy me down
7	unknown plus oil	Not as thick but clear

Explain your results:_____

_____

Procedure 1b: Grease Spot Test

Another test for the presence of lipids is the grease spot test. When grease is applied to a brown paper bag or paper towel, it leaves a translucent area. This area is easily distinguished from a wet spot left by water. This is also a very good test for the presence of a lipid in either solid or liquid form.

1. To a brown paper towel apply a drop of water. Mark the area.
2. To another area on the towel, apply a drop of oil. Mark the area.
3. Repeat step 2 for white bread and for the unknown lipid solution.
4. Allow all spots to dry for 15 minutes and note the changes. Record your results in Table 6.2.

Results

TABLE 6.2   Grease spot test.

Samples	Observations
oil	
water	
white bread	
unknown	

Explain your results:_____

_____

ACTIVITY 2: THE BIURET TEST FOR THE PRESENCE OF PROTEIN

Biuret reagent, which contains copper sulfate and sodium hydroxide, will react with a peptide bond to give a violet color. Free amino acids do not react with Biuret reagent; however, chains of more than six amino acids will react.

Materials

Reagents:
Biuret reagent

Test Solutions:
water, dilute gelatin
egg albumen, histidine, trypsin

unknown protein solution 1

Supplies:
test tubes
test tube racks
stirring rods
wax pencil or marker
1 mL pipettes
5 mL pipettes
pipette dispenser

Procedure

1. Label six test tubes 1 through 6.
2. To test tube 1, add 1 mL of water and 2 mL of Biuret reagent.
3. To test tube 2, add 1 mL of egg albumen and 2 mL of Biuret reagent.
4. To test tube 3, add 1 mL of histidine and 2 mL of Biuret reagent.
5. To test tube 4, add 1 mL of trypsin and 2 mL of Biuret reagent.
6. To test tube 5, add 1 mL of gelatin solution and 2 mL of Biuret reagent.
7. To test tube 6, add 1 mL of unknown protein solution 1 and 2 mL of Biuret reagent.
8. Allow each tube to stand for 5 minutes.
9. Observe and record your results in Table 6.3. For each test, indicate the color you observe, and whether this is a positive or negative result for the presence of protein.

Results

TABLE 6.3 Biuret test results.

Test Tube	Sample	Color of Solution	Protein? (yes/no)
1	water	Blue	No
2	egg albumen	Violet	Yes
3	histidine	Blue	No
4	trypsin	Blue	No
5	dilute gelatin	light violet/Purple	Yes
6	unknown protein 1		

ACTIVITY 3: DETERMINING PROTEIN CONCENTRATION

It is often important to determine the amount of protein that is present in a sample. This can be done by performing a protein **assay**. There are three types of protein assays that are in use at this time: the Folin assay, more commonly referred to as the Lowry assay; the Bradford assay; and the Biuret assay. All three methods will allow you to determine protein concentration. The only difference is in the sensitivity of each assay.

In this exercise, you will use the Biuret assay to create a standard curve. From that curve, you will determine the concentration of protein that is present in both known and unknown protein samples. Biuret reagent reacts with proteins to produce a violet color. The intensity of color is proportional to the concentration of protein that is present in the sample. Color intensity can be quantified by measuring the amount of light that passes through a protein solution when it is placed in a **spectrophotometer**. Please refer to Appendix 6 for the proper use of the spectrophotometer.

Students will construct a standard curve from which the concentration of any protein sample can be determined. This curve will be generated using six samples of the protein bovine serum albumin (BSA).

Figure 6.6 is an example of a standard curve for protein. This graph can be used to determine the protein concentration of any substance based on its absorbance value. You can also calculate the slope of the line by using the formula $y = mx + b$. The slope can then be used to determine the concentration of a protein without plotting it against the standard curve. Note that a standard curve need not be linear.

You will plot your standard curve on the graph provided in Figure 6.7. In addition, you will generate a standard curve using either Excel or another graphing program suggested by your instructor.

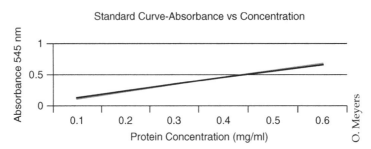

FIGURE 6.6 Standard curve.

Materials

Reagents:
12 mg/mL bovine serum albumin (BSA)
Biuret reagent

Test Solutions:
water, dilute gelatin
egg albumen

unknown protein solution 2
unknown protein solution 3
unknown protein solution 4

Supplies:
test tubes
test tube racks
wax pencil or marker
1 mL pipettes
5 mL pipettes
parafilm
spectrophotometer
cuvettes
holder for cuvettes
Kimwipes
vortex

Procedure

1. Refer to Appendix 6 for use of the spectrophotometer. Turn on the spectrophotometer and set the wavelength at 545 nm. Wait 15 minutes to allow the instrument to warm up before taking any readings.
2. Label six test tubes (numbers 1–6), making certain all test tubes are clean and dry. (To facilitate measuring, you can use capped graduated disposable tubes if available.)
3. To test tube 1, add 1 mL of water and 5 mL of Biuret reagent. This sample is called a **blank**.
4. To test tubes 2–6, add H_2O, BSA, and Biuret reagent *in that precise order*, according to Table 6.4.
5. Prepare test tubes 7–11 for your known and unknown protein samples by adding 1 mL of sample and 5 mL of Biuret reagent, according to Table 6.5.
6. For uncapped tubes, cover all 11 test tubes with parafilm and mix, either by using a vortex or by inversion. If using capped disposable tubes, there is no need for parafilm.
7. Arrange 11 cuvettes in order. **DO NOT MARK THE CUVETTES** as it will interfere with proper reading by the spectrophometer. Handle the cuvettes by the rim only. (See Appendix 6 for proper care of cuvettes.)
8. Pour the contents of each test tube into a separate cuvette.
9. Let tubes stand 5 minutes before reading in the spectrophotometer.
10. Before inserting into the spectrophotometer, wipe each cuvette with a Kimwipe or soft tissue to remove oils and fingerprints from the side of the tube.

TABLE 6.4 Bovine serum albumin (BSA) dilutions for the Standard Curve.

Test Tube	H_2O (mL)	12 mg/mL BSA (mL)	Biuret Reagent (mL)	Total mL	Final BSA Concentration (mg/mL)
1	1.0	0.0	5	6	0
2	0.9	0.1	5	6	0.2
3	0.8	0.2	5	6	0.4
4	0.7	0.3	5	6	0.6
5	0.6	0.4	5	6	0.8
6	0.5	0.5	5	6	1.0

TABLE 6.5 Setup of protein samples.

Test Tube	Protein Sample	Sample Amount (mL)	Biuret Reagent (mL)	Total mL
7	egg albumen	1	5	6
8	dilute gelatin	1	5	6
9	unknown 2	1	5	6
10	unknown 3	1	5	6
11	unknown 4	1	5	6

11. Place your blank (test tube 1) into the spectrophotometer and set the absorbance (ABS) at zero to calibrate (see Appendix 6). *Why do you need the blank?*

12. For each cuvette, place into the spectrophotometer and obtain an absorbance reading at 545 nm. Record the absorbance (ABS) in Table 6.6 below.

13. Using the absorbance measurements from samples 1–6 in the shaded area of Table 6.6, generate a standard curve graph in Figure 6.7. Plot *Protein Concentration* (given in Table 6.6) on the x-axis and *Absorbance (ABS)* on the y-axis. Label your axes properly and include appropriate units.

14. Use your standard curve to determine the concentration of protein in each of the known and unknown protein samples in test tubes 7–11. Record these results in Table 6.6.

15. Solutions must be removed from cuvettes as soon as possible to avoid staining. Clean the cuvettes with only soap and water. Rinse and return them to their proper place. **DO NOT USE A TEST TUBE BRUSH TO CLEAN A CUVETTE. IT WILL DAMAGE THE COATING.**

Results

TABLE 6.6 Absorbance and protein concentration.

Test Tube #	Protein Sample	ABS	Protein Concentration (mg/mL)
1	BSA	0.025	0
2	BSA	0.104	0.2
3	BSA	0.163	0.4
4	BSA	0.194	0.6
5	BSA	0.284	0.8
6	BSA		1.0
7	egg albumen	0.364	
8	dilute gelatin	0.310	
9	unknown 2	0.031	
10	unknown 3	0.028	
11	unknown 4	0.022	

Explain how you used your standard curve to find the protein concentrations for samples **7–11**. _____

_____
_____
_____

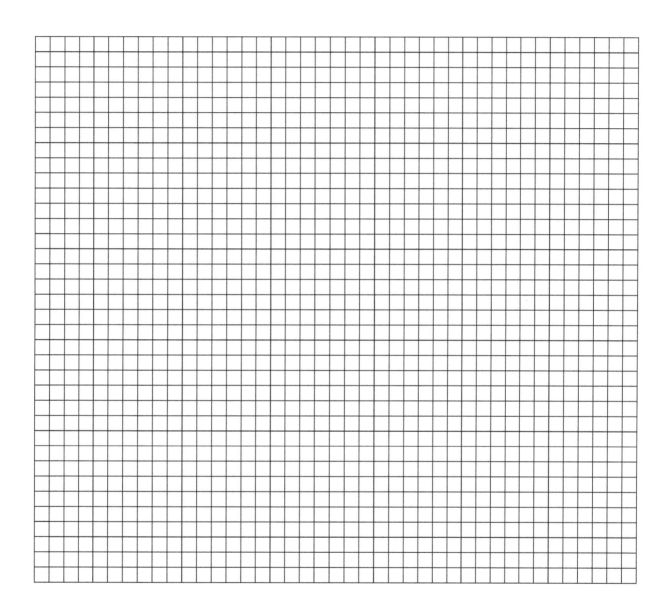

FIGURE 6.7 Standard curve for proteins.

III. COMPREHENSION CHECK

1. Sketch and describe the chemical structure of an amino acid.

2. What is the difference between an amino acid, a dipeptide, a polypeptide, and a protein?

3. What is a peptide bond?

4. What are the functions of proteins in living organisms?

5. What are the functions of lipids in living organisms?

6. What chemical property can be used to identify a lipid?

7. Which of the following substances will dissolve Sudan IV? Explain why or why not.
   a. vinegar
   b. margarine
   c. lemonade
   d. heavy cream
   e. skim milk

8. A Biuret assay was performed on an unknown protein solution to determine the total protein concentration. The absorbance (ABS) at 545 nm was 0.555. The standard curve for this assay is shown below. Estimate the concentration of protein in the unknown solution (include the proper units).

Answer: _____

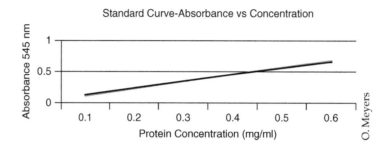

Standard Curve-Absorbance vs Concentration

Absorbance 545 nm

Protein Concentration (mg/ml)

O. Meyers

THE MICROSCOPE

After completing this laboratory exercise, students should be able to:

1. Properly use, transport, and care for a compound light microscope.
2. Comprehend the concepts of magnification and scale.
3. Identify and define the function of each part of the microscope.
4. Magnify a specimen and bring it into focus using the microscope.

I. INTRODUCTION

Similar to the way the telescope provides insight to explorers of the night sky, the microscope offers a window through which the biologist can study the cellular world. The unaided human eye is capable of observing only objects that are larger than 0.2 mm. To see smaller objects, they must be magnified, or enlarged. This laboratory exercise will take you through a number of activities that will introduce you to the microscope and the unseen world of the cell.

Different kinds of microscopes have been developed to study materials that are as small as an atom. The type of microscope that is used is determined by what object is to be observed. Figure 7.1 shows a comparative scale of common biological specimens and the way in which they can be seen. All microscopes have characteristic **magnification** and **resolving powers**. Whereas magnification refers to the scale of enlargement of an image, resolving power describes the ability of a microscope to distinguish fine details within an object. The **compound light microscope**, used in most introductory biology courses, can generally magnify images up to 1000–2000 times and has a resolving power of about 200 nm. The **dissecting microscope** does not provide good resolution, but it enables up to 40 fold magnification necessary for the dissection and manipulation of very small objects. Both the compound and the dissecting microscopes use visible light to distinguish images. The **transmission electron microscope (TEM)**, the **scanning electron microscope (SEM)**, and the **atomic force microscope (AFM)** use electrons to enlarge images up to 150,000–1,000,000 times and have resolving powers far greater than that of the compound light microscope.

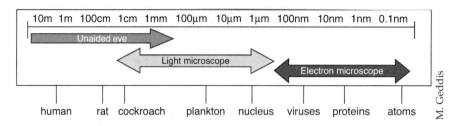

FIGURE 7.1 Scale of biological specimens.

The compound light microscope, shown in Figures 7.2 and 7.3, uses a combination of lenses to obtain its resolving power. The combined magnification power of an **ocular lens** and an **objective lens** allows us to distinguish objects as small as organelles, such as mitochondria. To determine the total magnification of a specimen viewed through a compound microscope, multiply the magnification strength of the ocular lens by the magnification strength of the objective lens. For example, if the ocular lens magnifies an image 10 times (10×) and the objective lens magnifies the image 40 times (40×), then the total magnification of the image is 10× multiplied by 40×, which equals 400×.

Practice Problems

1. With a total magnification of 400×, how small an object could you observe?

   Solution: $\dfrac{0.2 \text{ mm (resolution of the unaided eye)}}{400 \text{ (total magnification power)}} = 500 \text{ nm}$

2. Calculate the objective power needed to observe an object that is 1 μm in size.

   Answer: _____

II. LABORATORY ACTIVITIES
ACTIVITY 1: IDENTIFYING PARTS OF THE MICROSCOPE

Before using the microscope, it is necessary to identify its parts and become familiar with their functions.

Materials

Olympus C×31 compound light microscope

Procedure

1. Carefully carry a microscope from the microscope cabinet to your lab bench by holding one hand under the **base** and using your other hand to grasp the handle in the **arm** (refer to Figure 7.2). **ALWAYS USE TWO HANDS TO CARRY YOUR MICROSCOPE**.

2. Notice that all microscopes are numbered and are stored in a corresponding numbered space in the cabinet. You will use the same microscope for the duration of the semester. Record the number of your microscope here _____.

3. Familiarize yourself with the terms below and on the schematic diagram in Figure 7.2. Use each term to identify and label the parts of the microscope in Figure 7.3.

Parts of the compound light microscope:

ocular lenses: one or two eyepiece lenses that magnify a specimen by the factor engraved on the side of the eyepiece (usually 5× or 10×)

diopter adjustment ring: compensates for the difference in vision between the user's two eyes

interpupillary distance scale: measures the adjustable distance between the ocular lenses

objective lenses: a second set of lenses, each having a different magnification power; 4× (scanning lens), 10× (low power), 40× (high power) , and 100× (oil immersion lens)

rotating nosepiece: rotates the objective lenses

stage: the platform that supports the specimen to be observed

specimen holder: a metal support structure that secures a specimen slide in place

X-axis and Y-axis adjustment knobs: control knobs on a vertical bar that move the stage in the X or Y plane.

coarse focus knob: (inside larger knob): adjusts the stage position in large increments to bring a specimen into focus

fine focus knob (smaller outer knob): adjusts the stage position in small increments to sharpen the focus

pre-focusing dial: assures that the objective lens does not come into contact with the specimen slide

arm: upright bar connecting the ocular lens(es) to the rest of the microscope

light source: incandescent or fluorescent lamp that illuminates a specimen

on/off (I/O) switch: controls the light source *(I is "on", O is "off")*

light intensity control knob: adjusts the amount of light projected from the light source

condenser: lens system that concentrates the light coming from the light source and projects it through a specimen on the stage

field iris diaphragm: adjusts the diameter of the illuminated field of view

aperture iris diaphragm knob or lever: adjusts the amount of light passing through a specimen

base: a support that provides stability for the microscope

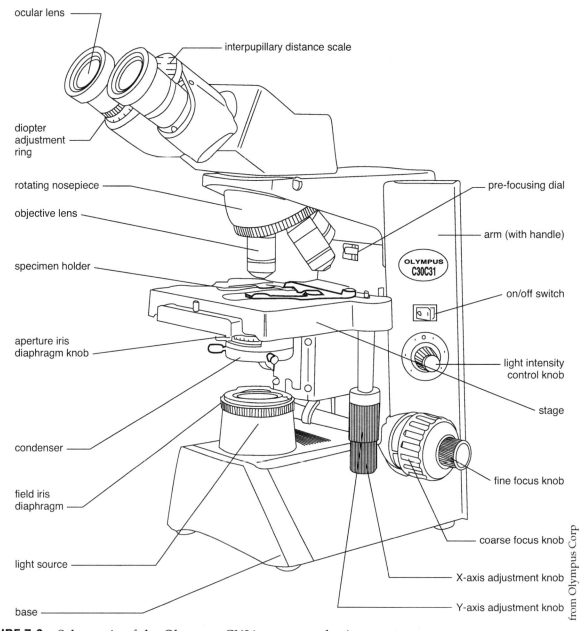

FIGURE 7.2 Schematic of the Olympus CX31 compound microscope.

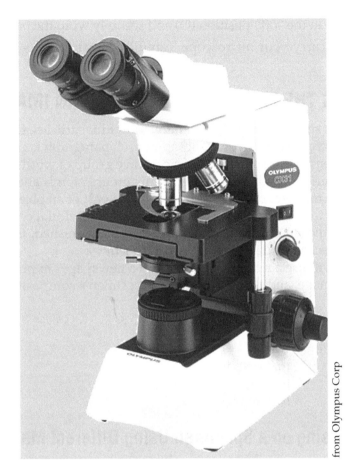

from Olympus Corp

FIGURE 7.3   Olympus CX31 compound microscope.

ACTIVITY 2: GENERAL CARE AND USE OF THE MICROSCOPE

The microscope is a sensitive and expensive piece of equipment, but with proper care it can serve as an indispensable biological tool for many years. You are responsible for proper care of your assigned microscope. The following procedure outlines general care instructions to follow each time you use your microscope in lab.

Materials

    compound microscope
    lens paper

Procedure

1. Plug the power cord into an outlet. Turn the light intensity control knob to 4 and open the field iris diaphragm. Press the light switch to the "on" position (I). The light source should illuminate.

2. Using **lens paper**, clean all lenses of your microscope. **USE LENS PAPER ONLY TO CLEAN YOUR LENSES. NEVER USE PAPER TOWELS OR KIMWIPES ON MICROSCOPE LENSES.**

3. Before storing the microscope, always remove slides from the stage.

4. If the stage is wet, carefully wipe it dry.

5. Be sure the lowest power objective lens (4×) is snapped into place for storage.

6. Press the light switch to the "off" position (O) and securely wrap the power cord around the base.

7. Using two hands to carry your microscope, carefully return it to its numbered location on the shelf.

ACTIVITY 3: USING THE MICROSCOPE TO VIEW AN IMAGE

Notice how an entire newspaper is made up of many individual articles. Each of these articles can be broken down into an ordered sequence of paragraphs; each paragraph is constructed from a group of related sentences; each sentence is comprised of a cohesive group of words; and, finally, each word is comprised of a specific series of individual letters. This progression is analogous to what a biologist observes when using the microscope to focus in on a small part of an object. At first, an entire specimen is viewed, such as a tissue sample *(a news article)*. Next, individual cells within the tissue can be seen *(the words)*. Finally, the smallest parts of cells can be distinguished, from organelles all the way down to the level of molecules and atoms *(the individual letters)*.

In this activity, you will use a single letter cut from a newspaper to practice how to magnify and focus on a microscopic image, and how to examine a specimen's composition and depth.

Materials

compound microscope	water
newspaper	razor blade
microscope slides	forceps
cover slips	hair strand

Procedure 3a: Focusing on a Specimen Using Different Magnifications

1. Obtain a local newspaper.

2. Using a razor blade, cut out a single letter (from an article, not a headline).

3. Prepare a microscope slide for a **wet mount** by placing 1 drop of water on the slide.

4. Using forceps, place your letter in the drop of water.

5. To complete the wet mount, cover the drop of water with a cover slip as shown in Figure 7.4. First, touch one edge of the cover slip to the drop of water and hold at an angle. Next, slowly let the cover slip drop onto the microscope slide. This will avoid trapping air bubbles under the cover slip.

6. Click the 4x scanning lens into place. Set the slide on the stage and secure it with the specimen holder. Turn on the light source and set the light intensity control at 4.

7. Use the stage control knobs (X and Y) to center the letter in the middle of the light, directly under the objective lens and over the aperture in the stage.

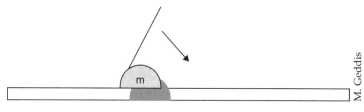

M. Geddis

FIGURE 7.4   Wet mount slide.

8. Looking from the side of the stage, use the coarse focus adjustment knob to move the stage in the Z plane (up and down). **DO NOT ALLOW THE OBJECTIVE LENS TO TOUCH THE SLIDE.** When the stage is within ½ inch to 1 inch from the objective lens, look through the ocular lenses. With a binocular microscope, you will need to adjust the interpupillary distance by moving the ocular lenses either closer together or further apart. You should see only one perfectly round circle. This is your **field of view**. Slowly move the stage down until the letter comes into view. The letter will appear as a fuzzy image.

9. Focus your image by looking through only one ocular lens at a time. Focus with your right eye by first turning the coarse adjustment knob, then the fine adjustment knob until you see a sharp image of your letter. Next, focus your left eye by turning the diopter adjustment ring on the left eyepiece. This will compensate for the difference in vision power between your two eyes. You should now see a sharp image of your letter when viewing with both eyes together.

10. While using the 4x scanning objective lens, center your letter again using the stage control knobs, then refocus the letter using only the fine adjustment knob. If necessary, you can adjust the amount of light coming through the specimen by shifting the aperture iris diaphragm lever. If the light is not bright enough, use the light intensity control knob to increase the amount of light projecting from the light source.

11. To increase the magnification of the specimen, use the higher power objective lenses. If the microscope is parfocal, the specimen will remain in focus as the magnification is increased. If a slight adjustment in focus is necessary, use only the fine adjustment knob.

12. View your letter under the scanning objective lens (4×) first, then with the low power lens (10×), and finally with the high power lens (40×). Before switching to a higher power lens, always be sure that your specimen is centered in your field of view.

13. Record your observations in Figure 7.5 by drawing exactly what you see in your field of view using each of the objective lenses indicated. *Do not use the 100x oil immersion lens.*

Results

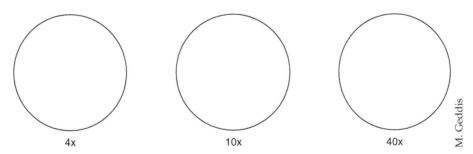

FIGURE 7.5   Drawings of a microscope specimen using different objective magnifications.

Procedure 3b: Depth of Focus

1. Remove a strand of your hair by gently plucking it from your scalp so that one end retains the hair follicle *(this is the part of the hair from which DNA is extracted for DNA testing).*

2. Use a razor blade to cut the strand of hair into two 1 cm long pieces.

3. Prepare a wet mount slide as described above. Cross the two pieces of the hair, one over the other.

4. Use the fine adjustment knob to focus through the depth of the two crossed pieces of hair. Determine which piece is on top.

5. Record your observations in Figure 7.6 using different magnifications.

Results

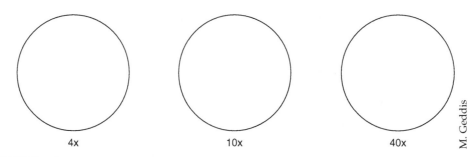

4x   10x   40x

M. Geddis

FIGURE 7.6   Drawings of cross hairs using different objective magnifications.

ACTIVITY 4: EXAMINING A SPECIMEN

When viewing a specimen using the light microscope, it is important to determine a suitable magnification to use. For example, to determine whether a sample of water from a spring or a well is potable, it can be examined using a microscope to see whether it is free of contaminants and microbes. Whereas one magnification power might be sufficient to examine the microscopic contents of the water sample, a lower magnification might fail to distinguish certain harmful impurities, such as bacteria.

Materials

compound microscope   pipettes
microscope slides   pond or river water
cover slips

Procedure

1. Obtain a sample of water from a pond or river (for example, the Hudson River).
2. Observe the specimen with your unaided eye and sketch what you see in Figure 7.7.
3. Prepare a wet mount of the water sample.
4. Observe the specimen using different magnifications and sketch what you see in Figure 7.7.

Results

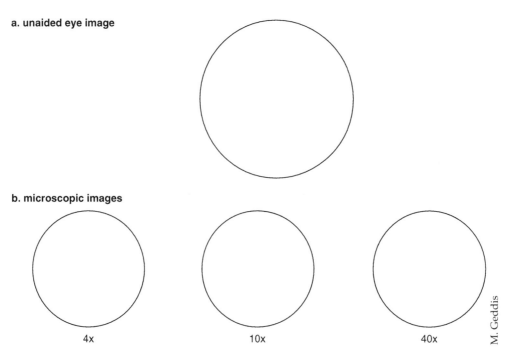

a. unaided eye image

b. microscopic images

4x   10x   40x

M. Geddis

FIGURE 7.7   Water samples as seen without (a) or with (b) the microscope.

ACTIVITY TO REINFORCE CONCEPTS
ACTIVITY 5: USING A DISSECTING MICROSCOPE

Often in biology, it is necessary to perform a task on a specimen. Although the object might be visible with the unaided eye, an increase in magnification could greatly enhance the efficiency of the task. For such procedures, the dissecting microscope is employed. A dissecting microscope has many of the same features as the compound light microscope, except that the resolving power is relatively poor, and the light source can be directed from either the top or the bottom. The advantage of this instrument is that a specimen can be manipulated while viewing under magnification.

Materials

dissecting microscope   pipettes
compound microscope   razor blade
slides   *Planaria* (flatworm)
cover slips

Procedure

1. Obtain a flatworm specimen and a dissecting microscope.

2. Observe the specimen with your unaided eye and describe your observations below.

3. Place the specimen on a microscope slide and observe under the dissecting microscope. Adjust the light source (above or below), the focus, and the objective magnification (0.7× to 4×). Remember that the overall magnification is the ocular lens magnification multiplied by the objective lens magnification. Record your overall magnification using the 4× objective lens: _____.

4. Adjust to a suitable magnification to view your flatworm. Record your overall magnification: _____.

5. Use a razor blade to transect the flatworm by cutting equidistant from the head and the tail *(a task that would be very difficult without the dissecting microscope)*.

6. Place a cover slip over the transected specimen and transfer to a compound microscope.

7. Observe the specimen using different magnifications and describe your results below.

8. Sketch your observations in Figure 7.8.

Results

Observational description with the unaided eye:

_____
_____
_____
_____

Observational description with the dissecting microscope:

_____
_____
_____
_____

Observational description with the compound light microscope:

_____
_____
_____
_____

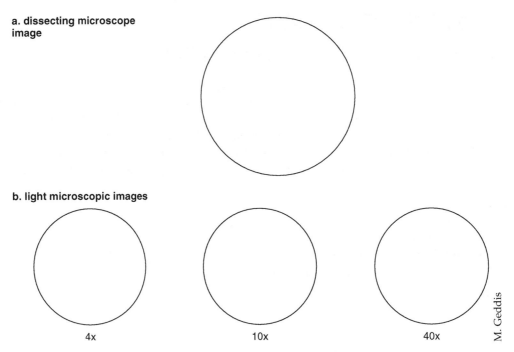

a. dissecting microscope image

b. light microscopic images

4x   10x   40x

M. Geddis

FIGURE 7.8 Images of a flatworm as viewed with (a) the dissecting microscope and (b) different objective magnifications of the compound light microscope.

III. COMPREHENSION CHECK

1. How many hands should be used to carry a microscope?

2. What factors determine the resolving power of a microscope?

3. Match the microscope that should be used to examine each different-sized object:

    _____ electron microscope   a. motor nerve
    _____ compound microscope   b. mitochondrion
    _____ dissecting microscope   c. plankton

4. What is the total magnification power of a compound microscope that has a 5× objective lens and 5× ocular lenses?

5. Why are binocular microscopes necessary for microscopy?

6. Explain the proper procedure for cleaning the lenses of the microscope.

PROKARYOTES AND EUKARYOTES

After completing this exercise, students should be able to do the following:
1. Distinguish the unique characteristics of prokaryotic and eukaryotic organisms.
2. Directly compare the size difference between a prokaryotic cell and a eukaryotic cell.
3. Prepare fixed and stained slides of organisms for microscopic examination.
4. Observe the process of phagocytosis by a unicellular eukaryote.

I. INTRODUCTION

The **cell** is the basic unit of structure and function that can perform all life activities. All organisms are composed of cells. Two different types of cells comprise all living things. **Prokaryotic** cells are those that lack an internal structure called a nucleus. Organisms that are composed of prokaryotic cells are known as **prokaryotes**. **Eukaryotic** cells contain a well-defined nucleus. Organisms that are made up of eukaryotic cells are known as **eukaryotes**.

A. Classifying Prokaryotes and Eukaryotes

All prokaryotic organisms are unicellular, and can be subdivided into two large classification groups or domains, namely **Domain Bacteria** and **Domain Archaea**. The Bacteria are the simplest organisms and include common bacteria such as *Escherichia coli* and *Staphylococcus*. The Archaea differ from the Bacteria in that they can survive in extremely inhospitable environments, including areas of unusually high temperature or conditions of either acidic or alkaline pH.

Eukaryotic organisms belong to a third domain, **Domain Eukarya**. All eukaryotes are further subdivided into four **kingdoms**: **Plantae**, **Animalia**, **Fungi**, and **Protista**. The kingdom Plantae is made up of all **autotrophic** multicellular plants. These organisms obtain nutrition by converting inorganic materials into organic compounds through the process of photosynthesis *(oak tree, onion)*. The Animalia, or animals, are multicellular **heterotrophs** that obtain nutrition by ingesting their food into a digestive opening *(tapeworm, lion, human)*. The Fungi are unicellular or multicellular eukaryotes that obtain nutrition by decomposing and absorbing nutrients from dead organic matter *(mushroom,* *bread mold).* The protists are a **paraphyletic** group, meaning they consist of those eukaryotes that are not animals, true fungi, or green plants. They are unicellular eukaryotes that share characteristics of all other kingdoms and have various modes of obtaining nutrients. There are many such protists and the relationships among these organisms are not as well defined as those of other kingdoms *(Amoeba, Paramecium).*

B. Comparing Prokaryotic and Eukaryotic Cells

The cells of prokaryotes and eukaryotes are similar in a number of aspects. For example:

1. Both contain DNA as their genetic material.
2. Both have a similar cell membrane structure composed of a phospholipid bilayer.
3. Both utilize structures called ribosomes for protein synthesis.
4. Both undergo similar metabolic processes to generate energy.
5. Both exhibit remarkable diversity in their forms.

Beyond these similarities, prokaryotic and eukaryotic cells differ considerably. Aside from the presence of a nucleus, eukaryotic cells are far more complex in structure, function, organization, and means of gene expression than are prokaryotic cells. Some characteristics that distinguish the two different cell types are summarized in Table 8.1.

TABLE 8.1 Comparison of prokaryotic and eukaryotic cells.

Characteristic	Prokaryotic Cells	Eukaryotic Cells
Nucleus	No nucleus is present.	A well defined nucleus is contained within each cell.
DNA	DNA is circular and is present within the cytoplasm of the cell in a region called a **nucleoid**.	DNA is linear. Along with proteins called histones, DNA is highly organized into **chromatin** within the nucleus of the cell.
Internal Structure	Prokaryotes lack complex membrane bound organelles.	Eukaryotic cells contain several membrane bound organelles, including mitochondria, chloroplasts, endoplasmic reticula, Golgi bodies, and vacuoles.
Cellularity	All prokaryotic organisms are unicellular, but can associate into clusters called colonies.	Eukaryotic organisms can be either unicellular or multicellular.
Size	On average, prokaryotic cells are much smaller in size than eukaryotic cells (1 μm to 10 μm).	On average, cells are much larger in size than prokaryotes (10 μm to 100 μm).
Reproduction	Prokaryotic organisms reproduce asexually by a process called binary fission.	Reproduction is asexual or sexual. Sexual reproduction involves a special type of cell division called meiosis to generate gametes.

II. LABORATORY ACTIVITIES

ACTIVITY 1: OBSERVATION OF PROKARYOTIC ORGANISMS

Bacteria are unicellular prokaryotes that range in size from 1.0 to 10.0 μm in length. They exhibit a wide variety of shapes including **spheres (cocci)**, **rods (bacilli)**, and **spirals**. Although bacteria are single-celled organisms, they often associate to form characteristic patterns such as diploids (*Neisseria*), chains (*Bacillus*, *Streptococcus*), or clusters of grapes (*Staphylococcus*).

All bacteria have an external cell wall made up of a chemical compound called **peptidoglycan**. One way in which bacteria can be differentiated from one another is by the thickness of their cell wall. A unique type of stain called **Gram stain** can be used to distinguish bacteria with a thick cell wall from those with a thin cell wall. Although bacteria lack membrane-bound organelles that are present in eukaryotes, recent studies have shown the presence of a bacterial cytoskeleton. Bacteria can be both heterotrophic and autotrophic. For example, the cyanobacteria can undergo the process of photosynthesis to produce their own food.

Although bacteria are commonly associated with disease, most bacteria are harmless. In fact, many bacteria are beneficial both to our health and the environment. Many bacteria play a major ecologic role by recycling nutrients back to the earth. Others serve an essential function in the food industry, for example by inhibiting the growth of harmful bacteria. Since the advent of biotechnology, bacteria have also been utilized by the pharmaceutical industry for manufacturing certain medications, such as human insulin used to treat type 1 diabetes.

Materials

light microscope	prepared slides of bacteria
clean slides	*Bacillus* cultures
cover slips	
small pipettes	

Procedure 1a: Prepared Slides of Prokaryotes

1. View the prepared slides of bacteria specimens using the 10x and 40x objective lenses.
2. Draw your observations in Figure 8.1 below.
3. Note the size and whether the bacteria you observe are spheres, rods, or spirals.

Procedure 1b: Wet Mount Slides of Live Bacteria

1. To prepare your wet mount, use a small pipette to place a drop of *Bacillus* culture onto a clean slide.
2. Carefully place a cover slip over the drop so that no air bubbles are trapped between the slide and the cover slip. (Refer to Laboratory Exercise 7 for how to prepare a wet mount slide.)
3. View your wet mount specimen using the 10x and 40x objective lenses.
4. In Figure 8.1 below, draw your observations under both low and high power.
5. Note how the live wet mount compares with the prepared slides from Procedure 1a.

ACTIVITY 2: OBSERVATION OF EUKARYOTIC ORGANISMS

Organisms of the domain Eukarya are divided into the kingdoms Plantae, Animalia, Fungi, and Protista. In this activity, you will examine specimens from each category.

The cells of eukaryotes generally range in size from 10 μm to 100 μm in length. Using the light microscope, more structures are visible in these cells than in the smaller and less complex bacterial cells. All eukaryotic cells contain a large nucleus that houses the DNA of the cell. In addition, many other cellular components can be seen that distinguish one cell type from another.

Plant cells are contained within a rigid cell wall composed of **cellulose**, different from the peptidoglycans in the cell wall of a bacterium. Plant cells also contain a large central vacuole that stores water and dissolved nutrients. Chloroplasts can also be observed within the cytoplasm of photosynthetic plant cells. The Fungi also contain a rigid cell wall, but it is composed of yet another compound known as **chitin**. Animal cells and most protists lack a cell wall, but can be seen enclosed in the phospholipid bilayer of the plasma membrane.

Freshwater protists contain specialized structures needed for survival in their habitats. Contractile vacuoles are present that serve to eliminate excess water that enters the cell of the organism. In addition, many protists have structural projections on the surface of the plasma membrane, such as cilia or flagella, that aid in movement through an aqueous environment. *Amoeba* have **pseudopodia**, extensions of the plasma membrane that function in movement and obtaining food particles. Under the light microscope, food vacuoles containing undigested food particles are easily observed.

Many other organelles are present within the cells of eukaryotes, but are difficult to distinguish clearly with the light microscope. These cellular components, such as endoplasmic reticula, Golgi bodies, lysosomes, and mitochondria, can be resolved clearly by electron microscopy.

Materials

light microscope	prepared slides of *amoeba* and *paramecia*
clean slides	*Elodea* leaves
cover slips	bread mold
small pipettes	live specimens of protists
	live nematode, *C. elegans*

Procedure 2a: Prepared Slides of Eukaryotes

1. View the prepared slides of *amoeba* and *paramecia* using the 10x and 40x objective lenses.

2. In Figure 8.2 below, draw your observations under both low and high power.

3. Note the differences in morphology between these organisms. Include all structures that you can identify and label your diagrams.

Procedure 2b: Wet Mount Slides of Eukaryotes

1. Prepare wet mounts of living samples of *Elodea*, bread mold, protists, and the nematode *C. elegans*. (Refer to Laboratory Exercise 7 for how to prepare a wet mount slide.)

2. View each wet mount specimen using the 10x and 40x objective lenses.

3. In Figure 8.3 below, draw your observations under both low and high power.

4. Note how a live wet mount of protists compares with the prepared slides you observed in Procedure 2a.

Results

L. Jayant

FIGURE 8.1 Microscopic images of bacteria.

L. Jayant

FIGURE 8.2 Microscopic images of protists on stained prepared slides.

L. Jayant

FIGURE 8.3   Microscopic images of live eukaryotes.

ACTIVITY 3: SIMPLE STAINING OF BACTERIA AND YEAST
Materials

light microscope   agar slants of *Bacillus* or *E. coli*
clean slides   agar slants of yeast
cover slips   crystal violet or methylene blue stain
wire inoculating loop   immersion oil
Bunsen burner   tongs

Procedure

1. On a clean slide, add a drop of water on both ends as shown in Figure 8.4.
2. Label each drop as B and Y underneath the slide.
3. Your instructor will demonstrate how to carefully light and use a Bunsen burner.
4. Sterilize an inoculating loop in non-luminous (blue) flame until it is red hot. **HOLD THE LOOP ONLY BY THE HANDLE. DO NOT TOUCH THE WIRE. IT WILL BE EXTREMELY HOT AND WILL CAUSE SEVERE BURNS ON YOUR SKIN.**
5. Cool the loop by touching it to the agar on the sides of the agar slant without touching the bacteria in the slant.
6. Once the loop is cooled, collect a small amount of the streaked bacteria from the agar slant onto the loop by gently scraping from the surface of the slant. Do not dig into the slant or you will pick up agar along with bacteria and will cause smudging on your slide.
7. Spread the bacteria into the drop of water labeled B.

FIGURE 8.4   Preparation of a slide sample for staining.

8. Sterilize the inoculating loop once again and cool as in steps 4 and 5 above.

9. Collect a small amount of yeast from an agar slant as in step 6 above.

10. Spread the yeast into the drop of water labeled Y.

11. Sterilize the inoculating loop once again as in step 4 above and place the loop upside down in a rack (the loop should be facing up).

12. Using tongs, carefully pass the slide in and out of the flame. This will heat fix the bacteria and yeast smears to the slide.

13. Once the smears are completely dry, cover the slide with crystal violet or methylene blue dye and allow to stand for 1 minute.

14. Rinse away the dye and air dry the slide. Do not wipe the slide dry.

15. **TURN OFF THE BUNSEN BURNER BY TURNING THE GAS VALVE TO THE OFF POSITION.**

16. The slide is ready to be viewed. View using the 10x, 40x, and 100x oil immersion objectives.

17. **THOROUGHLY CLEAN OIL FROM ALL MICROSCOPIC LENSES AND SLIDES. USE LENS PAPER ON LENSES, AND KIMWIPES ON MICROSCOPE HARDWARE AND SLIDES. YOUR INSTRUCTOR WILL EXAMINE YOUR MICROSCOPE BEFORE IT IS PUT AWAY.**

18. In Figure 8.5 below, draw your observations of the bacteria and yeast exactly as you see them under the microscope. Note the size difference between bacteria and yeast at all magnifications.

19. Compare your observations to the electron micrographs of prokaryotic and eukaryotic cells in Figure 8.6.

Results:

FIGURE 8.5   Microscopic images of stained samples of bacteria and yeast.

ACTIVITY TO REINFORCE CONCEPTS

ACTIVITY 4: OBSERVE PHAGOCYTOSIS IN THE PROTIST *TETRAHYMENA* (OPTIONAL)

Tetrahymena is a freshwater ciliated protist commonly used to study cell function, ciliary motion, cell proliferation, and **phagocytosis**, the process by which a cell engulfs solid particles and packages them into food vacuoles within the cell. When *Tetrahymena* encounter food, they use cilia to sweep food into an oral groove. This process can be visualized under the microscope as stained yeast cells are engulfed by the larger *Tetrahymena*. Students will actually count the number of stained food vacuoles formed within the *Tetrahymena* over a given period of time.

Materials

light microscope	test tube rack
clean slides	*Tetrahymena* grown on 2% protease peptone broth for 2 to 3 days
cover slips	Congo Red dye
small pipettes	methylcellulose
micropipette and tips	one package of active dry yeast
test tubes	

Procedure

1. Prepare a yeast suspension by adding 5 g of active dry yeast to 10 mL of lukewarm water.
2. Add 1 g of Congo Red to 10 mL of water in a test tube.
3. Add 1 mL of Congo Red solution to yeast suspension.
4. Using a small pipette, place 3 mL of *Tetrahymena* and 3 mL of yeast/Congo Red suspension in a test tube. Mix gently using a pipette.
5. Use a micropipette to add 10 µL of methylcellulose to a clean microscope slide.
6. Using a micropipette, remove 50 µL of *Tetrahymena*/yeast/Congo Red mixture and add directly to the methylcellulose on the slide. Exposing *Tetrahymena* to the methylcellulose will slow movement of the cells.
7. Place a cover slip over the sample and view using a light microscope.
8. Under the microscope, observe how the cells swim and form red food vacuoles by phagocytosis. Food vacuoles will appear as small red spots.
9. Count the number of red spots that are present in 10 cells. Record your results in Table 8.2.
10. Repeat steps 5 through 9 at 5, 10, and 15 minutes after adding *Tetrahymena* to the yeast/Congo Red mixture.
11. Using a small pipette, place 3 mL of *Tetrahymena* and 3 mL of Congo Red in a clean test tube (do not add yeast).
12. Repeat steps 5 through 9 at 0, 5, 10, and 15 minutes after adding *Tetrahymena* to the Congo Red mixture without yeast. Record your results in Table 8.3.
13. Discuss your results with your lab group. Explain whether *Tetrahymena* can distinguish between food and non-food materials.

Adapted from *The American Biology Teacher, 62(2): 136-139* by Donna M. Bozzone. Copyright © 1999 by The American Biology Teacher.

Results

TABLE 8.2 *Tetrahymena* with yeast.

Time (min)	0	5	10	20	30
Number of red spots					

TABLE 8.3 *Tetrahymena* without yeast.

Time (min)	0	5	10	20	30
Number of red spots					

Size and Structure Comparisons: Prokaryotic vs Eukaryotic Cells

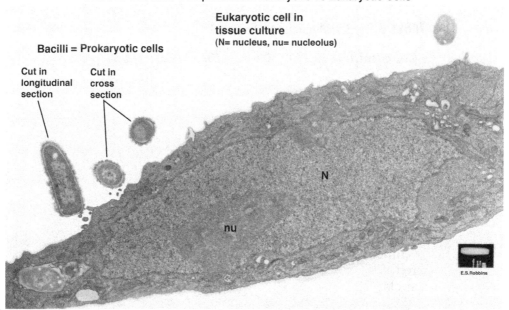

Transmission electron micrograph above, scanning electron micrograph below

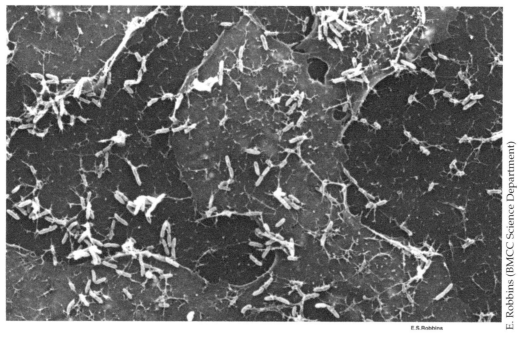

FIGURE 8.6 A comparison of prokaryotes and eukaryotes as seen in transmission and scanning electron micrographs. Note the difference in size. Courtesy of Edith S. Robbins, Ph.D.

III. COMPREHENSION CHECK

1. How are prokaryotic and eukaryotic cells the same? How are they different? Explain your answers.

2. Describe one example of a unicellular prokaryote and one example of a unicellular eukaryote.

3. What is the difference between a prepared (or fixed) slide and a wet mount slide? Explain their uses.

4. In the field of view below, draw what you might see in a light microscope at a total magnification of 800x if your slide contained both pond water protists and bacteria that were fixed and stained.

THE CELL

After completing this exercise, students should be able to do the following:

1. Distinguish between basic cell types.
2. Recognize specific cellular organelles.
3. Identify cellular organelles from microscope images.

I. INTRODUCTION

All living organisms are made of cells. From "simple" single-cellular organisms to "complicated" multicellular organisms, cells contain similar structures. Eukaryotic cells differ from prokaryotic cells because of the presence of a membrane-bound nucleus and membrane-bound organelles. Both plant and animal cells are eukaryotes, but plants differ by the presence of a cell wall and chloroplasts. It is the synergistic actions within and between cells that make life possible. Understanding cellular structure and function is essential for understanding biology.

II. LABORATORY ACTIVITIES

ACTIVITY 1. CELL TYPES

Materials

Table 9.1
Figures 9.1–9.4

Procedure

Using the list of organelles in Table 9.1 and labeled diagrams in Figures 9.1–9.3, familiarize yourself with different types of cells (prokaryotic, eukaryote plant cells, eukaryote animal cells) and organelles. Then, use this information to label and identify cell types and structures found in the micrographs shown in Figure 9.4.

TABLE 9.1   Cellular structures and descriptions.

Structure	Description
Nucleus	Contains DNA
Nucleolus	Darker region in nucleus, site of RNA transcription
Nucleoid	Region of circular DNA in prokaryote cells
Ribosome	Site of protein synthesis, directed by DNA; found in all cells
Smooth endoplasmic reticulum (sER)	Membrane structure responsible for processing newly synthesized products; synthesizes lipids
Rough endoplasmic reticulum (rER)	Membrane structure responsible for processing newly synthesized products; contains ribosomes
Mitochondrion	Powerhouse of the cell; produces ATP through oxidative phosphorylation
Chloroplast	Contains chlorophyll to convert light energy to chemical energy
Lysosome	Enzymatic digestion of cellular waste
Golgi apparatus	Membrane structure responsible for shipping cellular products
Plasma membrane	Phospholipid bilayer membrane that provides cell shape
Cell wall	Rigid outer wall of cellulose found in plants
Cytoplasm	Semi-fluid substance that contains the organelles inside the cell
Vacuole	Intercellular sac filled with fluid; used for storage of water and wastes

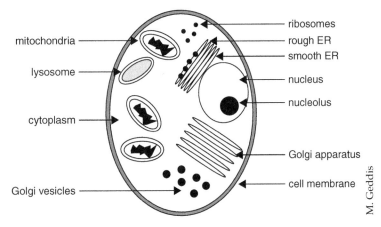

FIGURE 9.1   Animal cell.

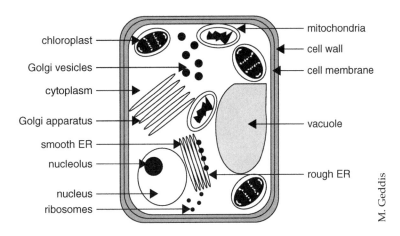

FIGURE 9.2   Plant cell.

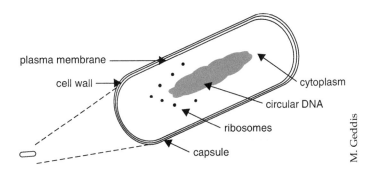

FIGURE 9.3   Prokaryotic cell.

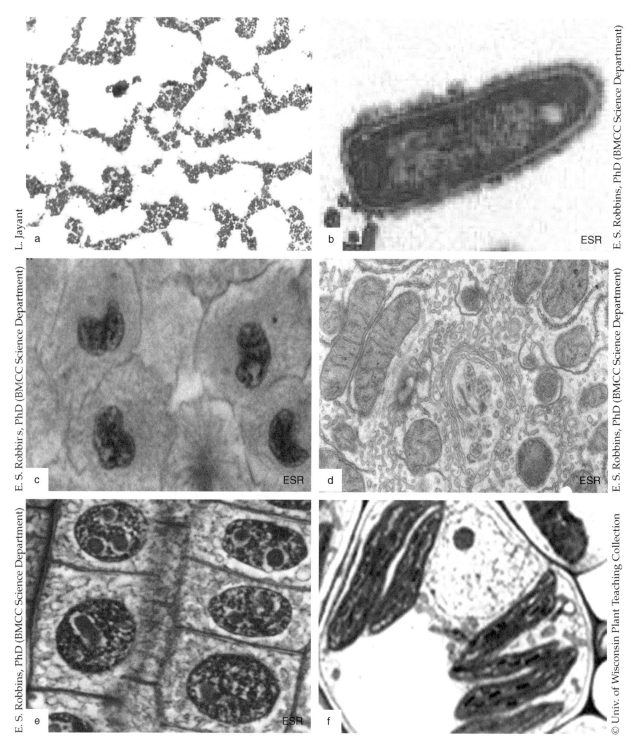

FIGURE 9.4   a, Light microscopic image of prokaryotes; b, Transmission electron micrograph of a prokaryote; c, Light microscopic image of animal cells; d, Electron micrograph of a small portion of the cytoplasm in an animal cell; e, Light microscopic image of plant cells; f. Electron micrograph of a portion of the cytoplasm of a plant cell.

Figures a-e courtesy of Edith S. Robbins, Ph.D.
Figure f © University of Wisconsin Plant Teaching Collection.

DIFFUSION AND OSMOSIS

After completing this exercise, students should be able to do the following:

1. Comprehend the concepts of diffusion, osmosis, and selective permeability.
2. Distinguish between hypertonic, isotonic, and hypotonic solutions.
3. Apply the concepts of diffusion and osmosis to living systems.
4. Determine the rate of a chemical process.

I. INTRODUCTION

A. Diffusion

Diffusion is defined as the spontaneous movement of solutes down their concentration gradient from an area of higher concentration to an area of lower concentration. It is the process of diffusion that drives the basic principles that allow life. The example illustrated in Figure 10.1 depicts free movement of solutes as they diffuse across a membrane. However, the plasma membrane of a living cell is actually **selectively permeable,** or semi-permeable, to many solutes. For example, small non-polar molecules such as hydrocarbons, CO_2, and O_2 can move freely through the hydrophobic center of the phospholipid membrane, but ions and polar molecules cannot. Cellular gates or channels embedded within the plasma membrane help to control the movement of such solutes into and out of the cell. By controlling what can and cannot move through the cell membrane, selective permeability leads to specialization for different types of cells.

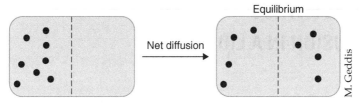

FIGURE 10.1 Diffusion down a concentration gradient.

The diffusion of solutes across a plasma membrane without the expenditure of energy is called **passive transport** and can occur with or without the aid of proteins. An example of this is **facilitated diffusion**, in which protein channels that span the plasma membrane allow ions and small polar molecules to bypass the hydrophobic portion of the phospholipid bilayer and diffuse into or out of the cell. Transport that requires the input of energy is referred to as **active transport**, in which solutes move against their concentration gradient from an area of lower concentration to an area of higher concentration.

B. Osmosis

Osmosis is a specialized type of facilitated diffusion that involves the selective transport of water molecules across a selectively permeable plasma membrane through protein channels called **aquaporins**. The process of osmosis acts to equilibrate the **osmolarity**, or relative concentration of solutes, outside the cell as compared to inside the cell. Given two solutions, the one with the higher solute concentration is said to be **hypertonic** to the second. Conversely, the solution with the lower solute concentration is **hypotonic** relative to the first. Two solutions that have the same concentration of solutes are at equilibrium and are said to be **isotonic** to one another.

Note that because dissolved solutes take up space in solution, a hypertonic solution will contain a lower concentration of water molecules relative to a hypotonic solution. Although water always continues to transport back and forth across a membrane in both directions, the net movement of water molecules will be from the hypotonic side to the hypertonic side, down the concentration gradient of water molecules, but against the concentration gradient of dissolved solutes.

C. Diffusion and Osmosis in Living Organisms

In all living cells, the properties of osmosis are necessary to maintain proper **osmotic pressure**, the pressure exerted on the cell membrane by the flow of water. Excess water movement into a cell will cause it to swell, whereas excess water movement out of a cell will cause it to shrink. Because either situation is potentially fatal to a cell, organisms have mechanisms that regulate water balance. For example, contractile vacuoles in unicellular fresh water protists serve to pump out excess water, and your kidneys function to remove excess water from your entire body. This type of water balance control is known as **osmoregulation**.

The principles that underlie the processes of diffusion and osmosis ultimately determine how individual living cells interact with their surrounding environment, which in turn influences the limits of cellular growth. For example, the transport of O_2 and glucose into a cell is required for cellular respiration, but is limited to the distance that these molecules can diffuse once inside the cell. Therefore, the surface area to volume ratio of an individual cell is essential for ensuring that solutes and molecules are equally distributed throughout the interior of the cell. Because the surface area to volume ratio of many small cells is far higher than that of one giant cell, the limits of diffusion compel cell division and multicellularity.

In this lab, you will examine the properties of diffusion through different kinds of substances, the biological significance of selective permeability of the plasma membrane, and how different cell types respond to changes in solute concentration.

II. LABORATORY ACTIVITIES
ACTIVITY 1: DIFFUSION IN A LIQUID
Materials

Celsius thermometer	hot plate	distilled water at room temperature
15 mL graduated cylinders (3)	ice	distilled ice water
stop watch	dye	distilled boiling water

Procedure

1. Add 10 mL of cold water to one cylinder, 10 mL of room temperature water to a second cylinder, and 10 mL of hot water to the third.
2. Record the temperature of each water sample in °C.
3. To each cylinder, add 1 drop of dye to the water. Do not mix the dye and water. Let the cylinders sit and observe the diffusion of the dye in each.
4. Use Table 10.1 to record the time in seconds that it takes for the water to turn the color of the dye uniformly.

Results

TABLE 10.1   Diffusion time of a liquid

Solution	Temp. (°C)	Time (sec)
Cold		
Room temperature		
Hot		

ACTIVITY 2: Diffusion in a Solid

Materials

agar plates   methylene blue crystals (MB)
forceps   potassium permanganate crystals
metric ruler

Procedure

1. Obtain one agar plate.
2. Place equal amounts of methylene blue ($C_{16}H_{18}N_3SCl$) crystals and potassium permanganate ($KMnO_4$) crystals approximately 70 mm apart on the agar surface and 10 mm from opposite edges of the plate as shown in Figure 10.2. Be sure to note which dye is which.
3. Use a metric ruler to measure the distance in millimeters (mm) that each dye diffuses at 15 minute intervals over a period of 90 minutes.
4. Record your results in Table 10.2.

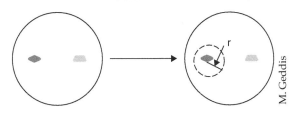

FIGURE 10.2   Diffusion of a solid.

Results

TABLE 10.2   Diffusion distance of a solid

Time (min)	MB (mm)	KMnO (mm)
15		
30		
45		
60		
75		
90		

ACTIVITY 3: DIFFUSION IN A GAS

Materials

ammonium
meter stick
stop watch

Procedure

1. Using the meter stick, measure the distance in meters (m) from the front of the room to your lab bench. Record this distance below.
2. Your instructor will open a bottle containing ammonium.
3. Using the stop watch, determine the time in seconds (sec) that it takes for you to smell the ammonium. Record this time below.
4. Using your distance and time measurements, calculate the rate of diffusion (m/sec) of a gas in air.

Results

Distance from front of room to your lab bench (m) = _____

Time to your lab bench (sec) = _____

Diffusion rate (m/s) = _____

ACTIVITY 4: DIFFUSION LIMITS OF CELLULAR EXCHANGE

The cellular membrane is the only means of contact with the external environment. This interface not only allows a constant exchange of nutrients and waste products, but it regulates the passage of substances critical for metabolic activity. Accordingly, there must be enough surface area to accommodate the increasing volume within a growing cell. This exercise will demonstrate how smaller cells are more efficient than larger ones in handling increased cellular volume.

Materials

metric ruler   paper towels
beaker   0.1 M NaOH
spoon   agar blocks of various sizes prepared with phenolphthalein
scalpel blade

Procedure

1. Record all measurements in Table 10.3 below.
2. Using a metric ruler, measure the length (l) in millimeters (mm) of one side of each agar block.
3. Calculate the surface area (mm^2) and volume (mm^3) of each cube using the following formulas:

Surface Area (S) of a cube = 6 × length squared (l^2)
Volume (V) of a cube = length cubed (l^3)

4. Calculate surface area to volume ratio (S/V). Record your answer.
5. Soak each cube in a beaker containing 0.1 M NaOH for 5 minutes.
6. Remove the agar cubes with a spoon and blot them dry with a paper towel.
7. Refer to Figure 10.3. On one side of each cube, mark the half-way point (dotted line). Measure the *half-length* distance (A) in mm from the half-way point to the edge of the cube. Record your measurements.
8. With your scalpel blade, cut each agar cube in half. Measure the depth of the pink-colored area (B) in mm and record it as *movement*. Calculate *movement* to *half-length ratio* (B/A). Record your measurements.
9. Use your measurements to construct a graph in Figure 10.4 showing B/A (movement/half-length ratio) as a function of S/V (surface area/volume ratio).

Results

TABLE 10.3   Diffusion-limited exchange

Length (L)	Surface area (S)	Volume (V)	S/V ratio	Half-length (A)	Movement (B)	B/A ratio

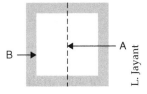

FIGURE 10.3   Diagram of an NaOH-soaked agar cube. A is the half length of the cube; B is the distance moved by the NaOH.

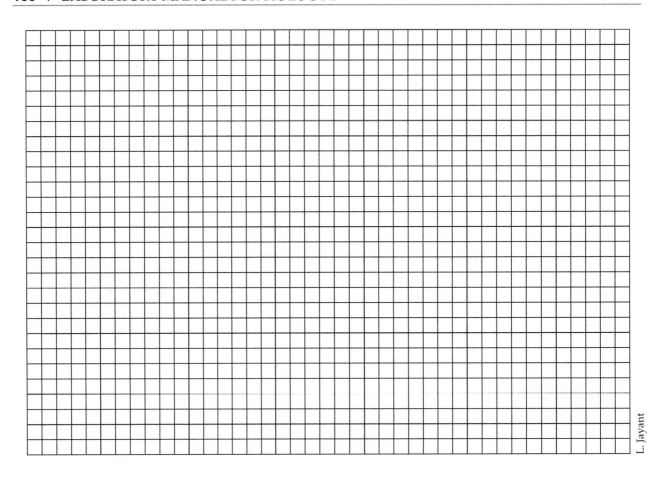

L. Jayant

FIGURE 10.4   Effect of increasing size of an object on the rate of diffusion.

ACTIVITY 5: OSMOSIS THROUGH A SELECTIVELY PERMEABLE MEMBRANE

Materials

dialysis tubing
twine
250 mL beakers (2)
test tubes (6)
test tube rack
marking pencil

pipettes
Benedict's solution
balance
40% maltose solution
distilled water

Procedure

1. Obtain four pieces of dialysis tubing, each 15 cm long, and presoak in distilled water.
2. For each piece of tubing, fold over one end. Tie the folded end securely with twine. This will make the tubing into a bag.
3. To two of the dialysis bags, insert 15 mL of 40% maltose solution into the open end.
4. To the other two bags, insert 15 mL of distilled water into the open end.

5. Fold and tie off the open end of each bag as in step 2, leaving only a small amount of air in each bag.

6. Observe the color of the solution within each bag. Record the weight in grams (g) of each bag and record in Table 10.4 in the column labeled "0 min." Rinse the outside of the bags with distilled water.

7. Add 100 mL of distilled water to a 250 mL beaker and add 100 mL of 40% maltose solution to another 250 mL beaker.

8. Into the beaker containing water, immerse one bag containing distilled water and one bag containing 40% maltose. Mark each bag clearly to indicate its contents. Make sure each bag has been rinsed before proceeding.

9. Into the beaker containing 40% maltose, immerse one bag containing distilled water and one bag containing 40% maltose.

10. After 10 minutes, remove each bag from its beaker, dry lightly on a paper towel, and weigh the bag in grams. After weighing, replace each bag in its beaker. Record the weight in Table 10.4.

11. Repeat step 10 at 10, 20, 30, 40 and 50 minutes. Record each weight in Table 10.4.

12. After the final weighing at 50 minutes, set up 4 clean test tubes. Using a clean pipette for each, remove 1 mL from each dialysis bag and transfer to separate test tubes. Label each tube appropriately as in Table 10.5. (Be careful not to spill any of the contents of the dialysis bags into the beakers.)

13. Set up 2 additional clean test tubes. Using a clean pipette for each, remove 1 mL from each beaker and transfer to separate test tubes. Label each tube appropriately as in Table 10.5.

14. To each test tube, add 5 mL of Benedict's solution and place in a boiling water bath for 5 minutes. Record the color of any precipitate that forms in Table 10.5. *To compare the colors, hold a piece of white paper behind the test tubes.*

15. Use the data in Table 10.4 to construct a graph in Figure 10.5. Put time (min) on the x-axis and weight (g) on the y-axis. Label each set of points and determine the rate of osmosis for each sample.

Results

TABLE 10.4   Osmosis through selectively permeable membrane.

Conditions	0 min	10 min	20 min	30 min	40 min	50 min
hypertonic — Water in bag; Maltose in beaker	14.7 g	13.8 g	13.2 g	12.4 g	11.9 g	
isotonic — Water in bag; Water in beaker	14.3 g	13.0 g	12.4 g	12.8 g	12.4 g	
hypotonic Maltose in bag; Water in beaker	16.6 g	17.5 g	18 g	19.5 g		
isotonic Maltose in bag; Maltose in beaker	15.2 g	15 g	14 g	13 g		

TABLE 10.5   Results of Benedict's test.

Test Sample	Color of Contents after Benedict's Test
Water from dialysis bag in water beaker	
Maltose from dialysis bag in water beaker	
Water from dialysis bag in maltose beaker	
Maltose from dialysis bag in maltose beaker	
Water from water beaker	
Maltose from maltose beaker	

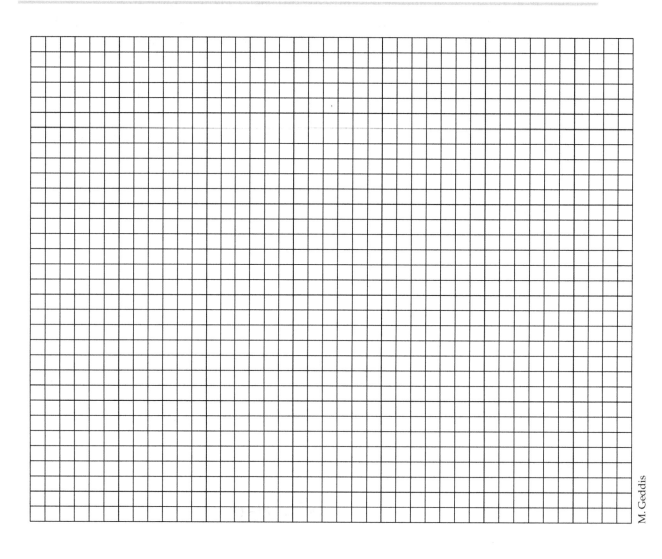

M. Geddis

FIGURE 10.5   Rate of osmosis of 40% maltose through a selectively permeable membrane.

ACTIVITIES TO REINFORCE CONCEPTS

ACTIVITY 6: OSMOSIS THROUGH A LIVING ANIMAL MEMBRANE

Materials

light microscope	animal blood
microscope slides (3)	distilled water (dH$_2$O)
cover slips	saline (0.9 % NaCl)
disposable gloves	15% NaCl solution

Procedure

1. Place 1 drop of animal blood on each of the three microscope slides.
2. To the first drop of blood, add one drop water; to the second drop, add one drop of saline; and to the third drop, add one drop of 15% NaCl.
3. After 5 minutes, place a cover slip over each sample to prepare wet mounts of the red blood cells.
4. Observe the red blood cells under the 40x objective lens of the microscope.
5. In Figure 10.6, draw and describe what you observe on each slide.
6. Discard the slides in an appropriate container.

Results

Water	Saline	15% NaCl

M. Geddis

FIGURE 10.6   Animal blood cells.

ACTIVITY 7: OSMOSIS THROUGH A LIVING PLANT MEMBRANE

Materials

light microscope	*Elodea*
microscope slides (3)	distilled water (dH$_2$O)
cover slips	saline (0.9 % NaCl)
	15% NaCl solution

Procedure

1. Place one leaf of *Elodea* on each of the three microscope slides.
2. To the first leaf sample, add one drop water; to the second sample, add one drop of saline; and to the third sample, add one drop of 15% NaCl.
3. After 5 minutes, place a cover slip over each sample to prepare wet mounts of the *Elodea* leaves.
4. Observe the *Elodea* leaf cells under the 40x objective lens of the microscope.
5. In Figure 10.7, draw and describe what you observe on each slide.
6. Discard the slides in an appropriate container.

Results

Water	Saline	15% NaCl

M. Geddis

FIGURE 10.7 *Elodea* cells.

ACTIVITY 8: EFFECT OF OSMOSIS ON A LIVING ORGANISM

Materials

dissecting microscope   flatworm (in pond water)
microscope slides (3)   distilled water (dH$_2$O)
cover slips   pond water
                            15% NaCl solution

Procedure

1. Obtain a flatworm in pond water and place it on a microscope slide.
2. Observe the flatworm under the dissecting scope. In Figure 10.8, draw and describe what you observe.
3. Carefully drain the pond water with a piece of paper towel and add one drop of 15% NaCl.
4. Observe the flatworm for several minutes. Draw and describe what you observe.
5. Carefully drain the NaCl with a piece of paper towel and add one drop of distilled water.
6. Observe the flatworm for several minutes. Draw and describe what you observe.
7. Discard the slides in an appropriate container.

Results

Water	15% NaCl

M. Geddis

FIGURE 10.8   Flatworm.

III. COMPREHENSION CHECK

1. How does temperature affect the rate of diffusion?

2. In your osmosis experiment, what do the results of the Benedict's test tell you?

3. Explain what would happen to each of the following cells if they were placed into a solution that is hypotonic, hypertonic, or isotonic:
   a. red blood cell –

   b. a cell from a spinach leaf –

   c. a fresh water amoeba –

4. You have two dialysis bags that are permeable to water and to NaCl, but not permeable to sucrose. You fill each bag with 10% NaCl and seal each end tight. One bag is to be placed in a solution of 10% NaCl. The second bag is to be placed into a solution of 10% NaCl and 30% sucrose. From your knowledge of diffusion and osmosis, explain what will happen to each as they sit in these solutions.

LABORATORY EXERCISE 11

ENZYMES

After completing these laboratory exercises students should be able to:

1. Comprehend the role of enzymes as biological catalysts.
2. Identify the substrate and product of an enzyme.
3. Plot a standard curve and use that curve to assay an enzyme quantitatively.

I. INTRODUCTION

An **enzyme** is a biological **catalyst** that speeds up a chemical reaction by reducing the **activation energy** (Figure 11.1). Most chemical catalysts are nonspecific, whereas an enzyme is very specific to the reaction it catalyzes. Enzymes reduce activation energy by binding to the reactants and by stressing certain bonds that need to be broken. Sometimes they briefly participate in the chemical reaction.

The reactants of an enzyme-catalyzed reaction are known as **substrates**. A substrate will bind to a specific location of an enzyme called the **active site.** The substrate binds to the active site through weak interactions, such as hydrogen bonds. The bound enzyme and substrate form what is known as an **induced fit**, an interaction in which the structure of the enzyme and substrate become slightly modified to fit together snugly. Enzymes are never consumed by a chemical reaction, but are recycled to be used again. Thus, very little enzyme is required to catalyze a reaction (Figure 11.2).

Some enzymes are made up of multiple parts, and all parts together are referred to as a **holoenzyme**. The protein component of a holoenzyme that contains the active site is called an **apoenzyme**.

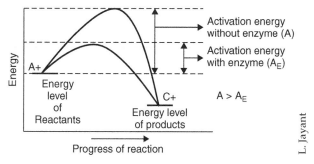

FIGURE 11.1   Activation energy in an enzyme–catalyzed and a non–catalyzed reaction.

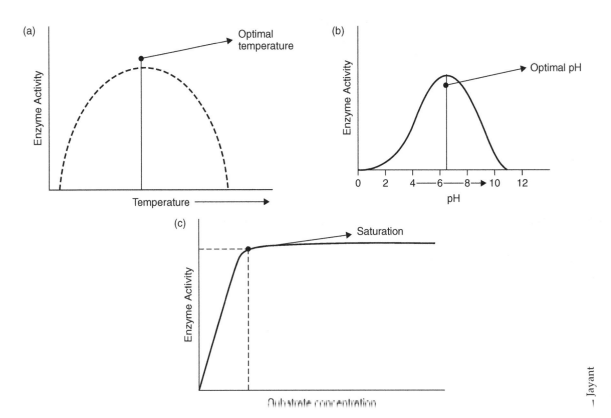

$$E+S \longrightarrow ES \rightarrow EP \rightarrow E+P$$

Enzyme is recycled

L. Jayant

FIGURE 11.2 S, substrate; E, enzyme; P, product. An enzyme is recycled after a product forms.

FIGURE 11.3 Effect of (a) temperature, (b) pH, and (c) substrate concentration on the rate of enzymatic activity.

Proper functioning of the holoenzyme often requires the association of an apoenzyme with either **cofactors** or **coenzymes**. A cofactor can be either a small inorganic molecule or an ion (such as Mg^{2+}), whereas a coenzyme is an organic compound (such as NAD^+ and biotin).

A. Factors That Influence Enzyme Activity

Like all proteins, the function of an enzyme depends on its three–dimensional conformation. Any factor that disrupts an enzyme's structure will inhibit catalytic function.

Temperature: Temperature has a critical effect on enzyme activity. Below an optimal level, as temperature increases, the rate of enzyme activity will also increase (Figure 11.3a). However, elevation beyond the optimal temperature will reduce the rate of enzymatic activity due to a breakdown of the enzyme's structure. The loss of conformation of a protein is called **denaturation**. All enzymes have an optimal temperature range within which they work best.

pH: All enzymes also have an optimal pH at which the enzyme can maintain its three–dimensional structure and have maximal activity. At a pH value too far above or too far below the optimal pH, the enzyme will become denatured and will no longer be active (Figure 11.3b).

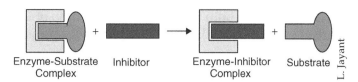

FIGURE 11.4   Enzyme inhibition.

Substrate concentration: Increasing the concentration of a substrate will increase the rate of enzyme activity, but only up to a certain level called a **saturation point**. At this point, the number of substrate molecules is equal to the number of enzyme molecules, and so all enzymes are bound to substrate. Adding more substrate beyond the saturation point will have no additional effect on the rate of enzyme activity (Figure 11.3c).

Enzyme inhibitors: Any factor that interferes with enzyme function is called an enzyme **inhibitor**. Temperature and pH are nonspecific inhibitors because they act by denaturing all proteins. Some inhibitors are more specific and prevent an enzyme from functioning by specifically binding to them. Specific inhibitors can be either **reversible** or **irreversible**. A reversible inhibitor can compete with a substrate for the active site of an enzyme. While the inhibitor is bound to the active site, the substrate cannot bind (Figure 11.4). The inhibition can be reversed if the substrate concentration is increased to a level at which it can out–compete the inhibitor for the active site on the enzyme.

Irreversible enzyme inhibitors bind to an enzyme in such a way that the enzyme is changed chemically and catalytic activity is permanently lost. Many poisons and toxins act by irreversibly inhibiting the action of important enzymes.

II. LABORATORY ACTIVITIES
ACTIVITY 1. PREPARATION OF STARCH

Starch is a polysaccharide found in many plants. It is made of glucose units held together by glycosidic bonds. Starch is made of two molecules: The linear amylose contains only α 1–4 glycosidic bonds and amylopectin contains α 1–4 and α 1–6 glycosidic bonds. Starch is a white odorless solid that is insoluble in water at room temperature.

Materials

conical flasks	thermometer
beakers	water bath set at 37°C
graduated cylinders	potato starch powder
pipettes	Lugol's iodine
glass plate	0.02 M sodium phosphate
hot plate	buffer (pH 6.9)

Procedure

1. Add approximately 40 mL of water to a large beaker and bring to a gentle boil.
2. Mix 1 g of soluble potato starch in approximately 5 mL of cold water.
3. While stirring, add the mixture to the boiling water.
4. Mix well and cool the starch solution to 37°C.
5. Add water to bring the total volume to 100 mL.

6. Put a few drops of the starch solution onto a glass plate. Add 1 drop of Lugol's iodine and see that a deep blue/black color develops. The blue/black color indicates the presence of starch in the solution.

7. Add 100 mL of 0.02 M sodium phosphate buffer (pH 6.9) to bring the volume to 200 mL.

Answer the following questions:
1. What percent starch solution did you prepare?_____
2. What is a buffer?_____
3. What kind of macromolecule is starch?_____

ACTIVITY 2. PREPARE STANDARD CURVE FOR MALTOSE

Maltose is a disaccharide composed of two monomer subunits of glucose held together by an α 1–4 glycosidic bond. This disaccharide is formed when **amylases** break down large starch polymers, for example, in germinating barley seeds. Maltose is sweet and is commonly used in the production of sweets or confections, such as candies and pastries. Like glucose, maltose has a free aldehyde (C=O) functional group, and hence is a reducing sugar. The standard curve will correlate maltose concentration (mg/mL) with absorbance (ABS) and will be used later in Activity 3 to determine the amount of maltose produced as starch is digested with amylase.

Answer the following questions:
1. What is a reducing sugar?_____
2. What test can you use to identify a reducing sugar?_____

Materials

beakers	cuvettes
graduated cylinder	pipettes (1 mL)
test tubes	pipette bulbs
test tube rack	kimwipes
hot plate	distilled water
spectrophotometer	dinitrosalicylic acid (DNS)
maltose	marking pencil

Procedure

1. Set up a boiling water bath.

2. Dissolve 200 mg of maltose in 50 mL distilled water, then bring the volume to 100 mL with distilled water. *What is the percent concentration of maltose in this solution?* _____

3. Label six test tubes at the top, 1 through 6. Set up the test tubes as shown in Table 11.1. Your instructor will demonstrate the proper use of pipette bulbs.

4. Place each test tube in a boiling water bath for 3 minutes. (Be sure to add DNS before boiling.)

5. Cool the tubes in ice for 5 minutes. → Ice wtr

6. Add 8 mL of distilled water to each tube. After diluting, calculate the concentration of maltose in mg/mL that is in each test tube. Record these concentrations in Table 11.1.

7. Using a spectrophotometer, read the absorbance of each sample at 540 nm (see Appendix 6). Use test tube 1 as your blank. Record your measurements in Table 11.1.

8. Use Figure 11.5 to plot a standard curve. Include units on your labels for each axis.

Your x axis will be _____.

Your Y axis will be _____.

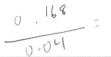

$$\frac{0.168}{0.04} =$$

Results

TABLE 11.1   Setup for maltose standard curve samples.

Test Tube #	1 (blank)	2	3	4	5	6
Water (mL)	1.0	0.8	0.6	0.4	0.2	0
Maltose (mL)	0	0.2	0.4	0.6	0.8	1.0
Dinitrosalicylic acid (mL) _before_	1	1	1	1	1	1
Water (after boiling) (mL)	8	8	8	8	8	8
Maltose concentration (mg/mL)	0	0.04	0.08	0.12	0.16	0.2
Absorbance (540 nm)	0	0.168A	0.328A	0.510A	0.814	0.992A

0.1

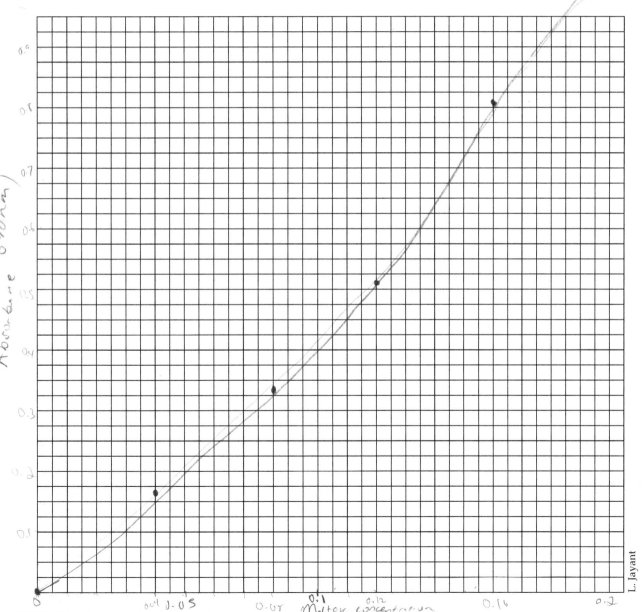

FIGURE 11.5   Standard curve for maltose.

L. Jayant

ACTIVITY 3. ENZYMATIC DEGRADATION OF STARCH BY α AMYLASE

The enzyme α–amylase catalyzes the hydrolysis of α 1–4 glycosidic linkages in polysaccharides that contain three or more D–glucose units. This enzyme randomly hydrolyzes glycosidic bonds in starch and glycogen, initially yielding sugars called dextrins, maltodextrins, and finally maltose and glucose.

Degradation of starch to maltose is an exergonic reaction and should proceed spontaneously. However, starches that are obtained from various food sources do not automatically break down to maltose. This is because _____
_____

An enzyme added to the starch solution will lower the _____ so that the reaction can proceed.

Maltose reduces dinitrosalicylic acid to a colored compound that can be detected and quantitated by a spectrophotometry at 540 nm. The reaction can be summerized as:

$$\text{Starch} \xrightarrow{\text{Alpha amylase}} \text{Maltose}$$
$$\text{Maltose + dinitrosalicylic acid} \longrightarrow \text{Colored product}$$

Answer the following questions:

1. What are the substrate, product, and enzyme in the above reaction? _____
_____
_____

2. Does the enzyme get used up as the reaction proceeds? Explain your answer. _____
_____
_____

3. Predict what should happen to the quantities of maltose and starch as the reaction proceeds after adding the enzyme. _____
_____
_____

Materials

test tubes	hot plate
test tube rack	spectrophotometer
conical flask	cuvettes
beakers	Kimwipes
pipettes (1 mL, 5 mL)	starch solution
pipette bulbs	distilled water
water bath	dinitrosalicylic acid (DNS)
timer	fungal alpha amylase

Procedure

1. Turn on the spectrophotometer and set at 540 nm (see Appendix 6).
2. Label two 250 mL conical flasks as '+ enzyme' and '– enzyme'. To each flask, add 100 mL of the starch solution prepared in Activity 1, and place them in a water bath set at 37 °C.

Adapted from Worthington, Von (Editor) 1993, *Worthington Enzyme Manual: Enzymes and Related Biochemicals*, Freehold, NJ.

3. Use two test tube racks and put seven test tubes in each rack.

4. Label the first set of tubes 1 through 7. This will be the "+ enzyme" set. (Samples of the "+ enzyme" reaction will be added to these tubes at 5 minute intervals during the experiment).

5. Label the second test of tubes −1 through −7. This will be the "− enzyme" set. (Samples of the "− enzyme" reaction will be added to these tubes at 5 minute intervals during the experiment).

6. To each of the 14 test tubes, add 1 mL DNS.

7. To prepare spectrophotometer blanks for the + enzyme and − enzyme sets, add 1 mL of 1% starch to test tube 1 and to −1. Mix each thoroughly and set these tubes aside at 37 °C.

8. Prepare fresh amylase by mixing 0.2 g amylase with 100 mL distilled water.

9. To the conical flask labeled + enzyme, add 5 mL freshly prepared amylase from step 8.

10. To the conical flask labeled − enzyme, add 5 mL distilled water.

11. After 5 minutes, remove 1 mL of the + enzyme reaction and add to test tube +2, and remove 1 mL of the − enzyme reaction and add to test tube −2. (The DNS in these test tubes will destroy any enzyme in the reaction).

12. Continue taking samples of the + enzyme and − enzyme reactions after each additional 5 minute interval. Add samples to corresponding test tubes labeled +3 through +7 (for the + enzyme reaction) and test tubes labeled −3 through −7 (for the − enzyme reaction).

13. After the 30 minute period, boil all tubes at the same time for 5 minutes.

14. Cool each tube in the 37 °C water bath.

15. Add 8 mL distilled water at 37 °C to each tube and mix well.

16. For the + enzyme set, use tube +1 to blank the spectrophotometer.

17. For tubes +2 through +7, measure the absorbance. Record your measurement in Table 11.2. From the Standard Curve in Activity 2, determine the concentration of maltose for each sample.

18. For the − enzyme set, use tube −1 to blank the spectrophotometer.

19. For tubes −2 through −7, measure the absorbance. Record your measurement in Table 11.3. From the Standard Curve in Activity 2, determine the concentration of maltose for each sample.

20. In Figure 11.6, plot a line graph of maltose concentration (mg/mL) vs. time (min.). Plot your data for both "+ enzyme" and "− enzyme" samples.

Results

Slope standard curve 4.2

TABLE 11.2 Absorption data for starch treated with amylase.

Tube #	1	2	3	4	5	6	7
Absorbance	0	2.40	2.484	2.533	2.547	2.513	2.594
Maltose (mg/mL)		0.526	0.591	0.603	0.606	0.598	0.6176

TABLE 11.3 Absorption data for starch not treated with amylase.

Tube #	−1	−2	−3	−4	−5	−6	−7
Absorbance	0.	0.077	0.076	0.072	0.071	0.070	0.070
Maltose (mg/mL)							

Answer the following questions:

1. What is the need for a flask with no enzyme? It is the _____
2. Explain why is it important to use the flask with no enzyme.

   _____

3. What is your dependent variable in this experiment?_____
4. What is your independent variable in this experiment?_____
5. What variable will be on the x−axis? _____
6. What variable will be on the y−axis? _____
7. Why did you add 5 mL of water to the − enzyme flask?_____

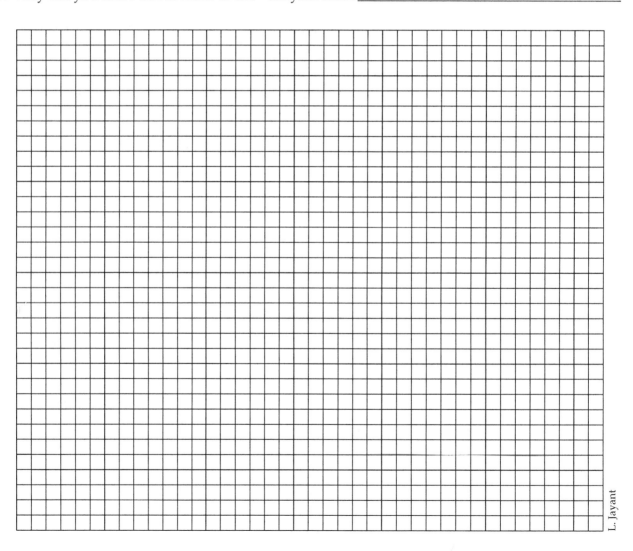

FIGURE 11.6 Maltose produced with and without enzymatic digestion of starch.

II. COMPREHENSION CHECK

1. The active site of an enzyme is the region that binds to the _____.

2. Succinate dehydrogenase normally catalyses the conversion of succinate to fumarate. If malic acid is added to this reaction, fumarate is no longer produced. Identify the enzyme, product, substrate, and inhibitor in this reaction.

3. A chemical reaction will proceed twice as fast for every 10°C rise in temperature. However, a reaction catalyzed by an enzyme will speed up only to a certain temperature at which point the reaction will cease. How do you explain this observation?

4. L–amino acid dehydrogenase is an enzyme that can catalyze the oxidation of different L–amino acids. It cannot catalyze the oxidation of D–amino acids or other L–compounds. This enzyme is very_____ for its substrate.

5. Not much enzyme is needed to catalyze a reaction. Why?

6. Most enzymes lose their function if they lose their structure. Most enzymes are also proteins. What type of structure do enzymes normally have?

7. A catalyst is a substance that
   a. increases the rate of a chemical reaction but is itself unchanged by the reaction
   b. is toxic to most cells.
   c. is composed mostly of lipids.
   d. does not usually participate in any chemical reactions.

8. A change in environmental pH can inhibit the proper functioning of an enzyme by
   a. changing the shape of its active site.
   b. altering the enzyme`s primary structure.
   c. altering the specificity of the enzyme.
   d. a, b, and c.
   e. a and c.

PHOTOSYNTHESIS

After completing this exercise, students should be able to do the following:

1. Name the plant pigments responsible for photosynthesis and identify some of these pigments using paper chromatography.
2. Comprehend the roles played by light and pigments during the process of photosynthesis.
3. Use a spectrophotometer to measure the absorption of different wavelengths of light by plant pigments.
4. Compare and contrast the action spectrum of photosynthesis with the absorption spectrum of plant pigments.

I. INTRODUCTION

Photosynthesis is the process by which all plants, as well as some bacteria and protists, convert solar energy into chemical energy and stores it as sugar. This fundamental biological process is crucial for survival of all life on earth. Either directly or indirectly, the food supply for all living organisms originates with photosynthesis.

Photosynthesis can be described as a series of chemical reactions in which carbon dioxide (CO_2) is converted into carbohydrates as a food source, and oxygen gas (O_2) as a by-product. All aerobic organisms, including plants and animals, require oxygen to generate ATP during the process of cellular respiration. Furthermore, the removal of carbon dioxide, a greenhouse gas, from the atmosphere is essential in the prevention of global warming. Thus, not only does photosynthesis provide both food and oxygen for organisms, it also helps to regulate the composition of our atmosphere.

All of the chemical reactions that occur during photosynthesis can be summarized as follows:

$$6\,H_2O \quad + \quad 6\,CO_2 \quad \xrightarrow{\text{light energy}} \quad C_6H_{12}O_6 \quad + \quad 6O_2$$

$$\text{\textit{water} \qquad \textit{carbon dioxide} \qquad\qquad \textit{glucose} \qquad \textit{oxygen}}$$

Six molecules of water plus six molecules of carbon dioxide, along with the input of light energy, yield one molecule of glucose plus six molecules of oxygen (which is released into the atmosphere).

II. LABORATORY ACTIVITIES

ACTIVITY 1: IDENTIFICATION OF PHOTOSYNTHETIC PIGMENTS BY PAPER CHROMATOGRAPHY

In plants, the entire process of photosynthesis takes place in specialized organelles called **chloroplasts.** Chloroplasts contain **chlorophyll,** the most important photosynthetic **pigment.** Other pigments found in chloroplasts are **carotenoids** and **phycobilins:**

1. **Chlorophyll:** This pigment gives plant leaves their green color. A molecule of chlorophyll consists of a porphyrin ring containing magnesium at the center (Figure 12.1). There are many types of chlorophyll molecules, including chlorophyll a, the primary photosynthetic pigment in green plants, as well as chlorophylls b, c, and d. Chlorophyll is insoluble in water and remains membrane-bound within the grana of the chloroplast.

2. **Carotenoids:** These pigments can be red, orange, or yellow in color. Carotenoid molecules are composed of two small six-carbon rings connected by a chain of carbon atoms. These compounds are also insoluble in water and therefore membrane-bound. There are two types of carotenoids:

   a. Hydrocarbon carotenoids are called **carotenes** *(examples are β-carotene, α-carotene, and lycopene).*

   b. Oxygenated hydrocarbon carotenoids are called **xanthophylls** *(examples are lutein and fucoxanthin).*

3. **Phycobilins:** These are water-soluble pigments found in the **stroma** of the chloroplast *(examples are phycocyanin, a blue pigment, and phycoerythrin, a red pigment).*

Many of these plant pigments can be separated from each other by the simple process of **paper chromatography.** The pigments are extracted from plant leaves and are placed on a strip of **chromatography paper.** As an organic solvent is pulled through the paper by capillary action, it carries the pigments along with it. The basis for separation is twofold: Each pigment will move at a different rate depending on (1) its solubility in the solvent and (2) its degree of attraction to the paper.

Chromatography paper contains cellulose and hence has polar properties. The organic solvent, however, is non-polar. Non-polar pigments, such as the carotenoids, remain dissolved in the organic solvent longer in comparison to polar pigments. Polar pigments, on the other hand, are more easily attracted to the paper. As a result, some pigments will travel faster than others through the

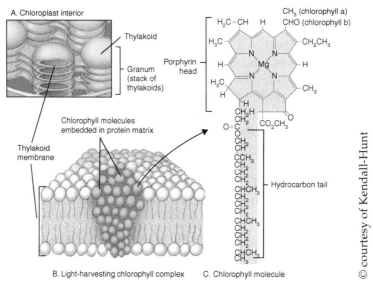

FIGURE 12.1 Chlorophyll.

chromatography paper, thereby separating along the way. The relative separation of each pigment can be compared by measuring what is known as the **Rf** value *(or retention factor),* which relates the distance traveled by a pigment to the distance traveled by the solvent during the chromatography process.

In this exercise, you will separate pigments extracted from spinach leaves and calculate the Rf value for each.

Materials

Whatman No 1 chromatography filter paper cut into strips
spinach leaf pigments extracted in acetone
capillary tube
chromatography solvent of petroleum ether and acetone
ventilation hood

large test tubes
rubber stoppers with hooks
pencil
forceps

Procedure

1. Cut a strip of chromatography paper the length of a test tube. Cut the bottom end of the strip to a point.

2. Use a pencil to mark a line 1 cm from the bottom end of the paper. *(Why not use a pen?)*

3. Using a capillary tube, apply a drop of the spinach leaf pigment extract on the pencil line (see Figure 12.2.) Allow the extract to dry. Repeat this procedure eight to ten times to provide a sufficient amount of extract in the spot. Hang the strip from the hook of a rubber stopper.

4. Working under a ventilation hood, fill one-tenth of a test tube with the solvent. Re-cap the solvent bottle. *Caution: Avoid inhaling the solvent.*

5. Carefully lower the strip, pointed end down, into the solvent, making sure that the spot is above the level of the solvent. The stopper should sit loosely on top of the test tube.

6. Let the setup stand until the solvent front has moved to within 3 cm of the top of the paper (see Figure 12.2.)

7. Remove the paper from the test tube and mark the solvent front. Let the paper air dry.

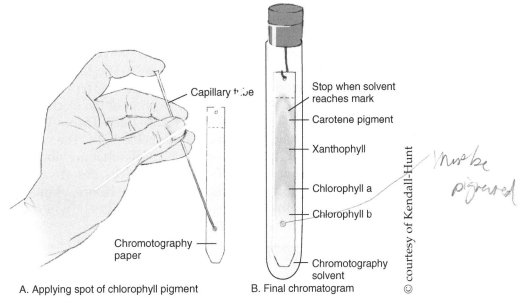

A. Applying spot of chlorophyll pigment   B. Final chromatogram

FIGURE 12.2   Setup for paper chromatography. The chromatogram in the tube shows the separation of photosynthetic pigments.

TABLE 12.1 Rf values of different photosynthetic pigments.

Pigment	distance P	distance S	Rf

8. On the dried chromatography paper, measure the distance traveled by each pigment (distance P) from the original line to the leading edge of the pigment.

9. Measure the distance traveled by the solvent (distance S) in the same way. Record your data in Table 12.1.

Results

1. In your notebook, diagram the appearance of the chromatography paper before and after the procedure. Identify the various pigment bands that separated and their colors.

2. Using the measurements you obtained in steps 8 and 9 of the procedure, calculate the Rf value for each pigment using the following formula and record in Table 12.1:

$$Rf = \frac{\text{Distance traveled by the pigment (P)}}{\text{Distance traveled by the solvent (S)}}$$

Discussion

Discuss with your group members what the Rf values mean with respect to polarity of the different pigments.

ACTIVITY 2: THE ACTION SPECTRUM AND THE ABSORPTION SPECTRUM

The **electromagnetic spectrum** is a continuum of all electromagnetic waves arranged according to frequency and wavelength. The sun, earth, and other planetary bodies radiate electromagnetic energy of varying wavelengths.

The spectrum of waves is divided into sections based on wavelength. The shortest waves comprise gamma radiation, with wavelengths of 10^{-3} nm or less. The longest waves are radio waves, with wavelengths of several kilometers. **Visible light** is a particular type of radiation that the human eye can see. Visible light is contained within a narrow portion of the electromagnetic spectrum that ranges from wavelengths of about 400 nm (violet) to 700 nm (red) (Figure 12.3).

Pigments selectively absorb certain wavelengths of visible light. The color of a pigment comes from the wavelengths of **reflected light,** or those wavelengths that are not absorbed. For example, chlorophyll, the green pigment common to all photosynthetic cells, absorbs all wavelengths of visible light except that for the color green (510 nm), which is reflected. Hence, plants appear to us as green. Black pigments absorb all wavelengths that strike them. Alternatively, white and very light-colored pigments reflect almost all wavelengths of light.

Every pigment has its own characteristic **absorption spectrum** with its own relative **absorbance** at different wavelengths of light. An absorption spectrum for a pigment can be determined using an instrument known as a **spectrophotometer** (see Appendix 6). In the process of spectrophotometry, a beam of light having a particular wavelength is passed through a solution. For each given wavelength, the spectrophotometer measures the relative proportion of light transmitted versus light absorbed by the pigments in the solution.

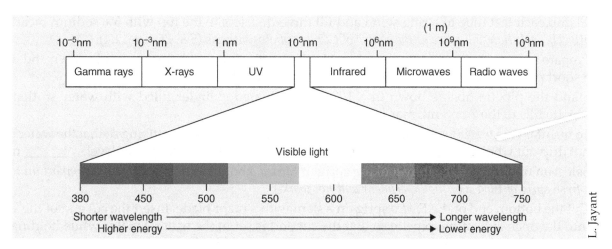

FIGURE 12.3   The electromagnetic spectrum.

When a pigment absorbs light energy, one of three things could happen: (1) The absorbed energy could be released as heat; (2) the absorbed energy could be immediately emitted as a longer wavelength *(a phenomenon known as fluorescence)*, or (3) the absorbed energy could trigger a chemical reaction, as in photosynthesis. Within the chloroplast of a photosynthetic cell, when light energy is absorbed by chlorophyll, a chemical reaction is sparked that begins the process of photosynthesis. Only wavelengths of light that are absorbed can initiate photosynthesis.

Whereas an absorption spectrum describes different wavelengths of absorbed light, an **action spectrum** for a photosynthetic pigment describes how effective different wavelengths of light are in initiating photosynthesis. An action spectrum for photosynthesis can be determined by measuring the rate of photosynthesis using different wavelengths of light.

In this exercise, you will:

a. determine the action spectrum for photosynthetic pigments by measuring the rate of photosynthesis at different wavelengths of light. *(Different wavelengths of light can be obtained by placing different colored filters in front of a light source. Because oxygen is a by-product of photosynthesis, the rate of photosynthesis can be determined by measuring the amount of oxygen released per minute.)*

b. determine the absorption spectrum for photosynthetic pigments by using a spectrophotometer to measure absorbance at different wavelengths of light. *(Refer to Appendix 6 for how to use a spectrophotometer.)*

Materials

Elodea leaves (that have been kept in the dark*)	beakers	100 mL graduated cylinder
spinach leaves	7 large test tubes	graduated pipettes with rubber tubings
blender	5% sodium	light source
cheesecloth	bicarbonate	ring stands and clamps (large, small)
spectrophotometer	Kimwipes	cellophane paper (red, yellow, green,
cuvettes	razor blade	blue, and violet)
	tape	

Procedure 2a: The Action Spectrum for Photosynthesis

1. *Place *Elodea* sprigs in the dark overnight and keep in the dark until ready to use.

2. Prepare five colored setups by wrapping each of five large test tubes in a different color cellophane paper. Tape the cellophane in place. Leave two additional test tubes unwrapped. *Each lab group will work with a different color setup.*

3. Clamp each test tube to a ring stand and fill three-quarters to the top with 5% sodium bicarbonate. *This solution will serve as the source of CO$_2$ for photosynthesis* (See Figure 12.4).

4. Prepare a separate pipette for each setup: Hold a 10 mL graduated pipette upside down and place a short piece of rubber tubing over the delivery (pointed) end.

5. Stand the pipette upside down in a 100 mL graduated cylinder filled with water so that the pipette fills to the 7 or 8 mL mark.

6. To maintain the water level in the cylinder, tightly attach a clamp to the tubing so that the water does not drip out when the pipette is removed from the cylinder. Record the water level: _____ mL.

7. Select an undamaged, healthy looking sprig of *Elodea*, about 15 cm in length. *(It is important to use a fresh sprig of Elodea or the experiment will not work.)*

8. Cut the bottom end of the *Elodea* sprig on a slant with a razor blade. Insert the cut end of the sprig into the open side of the pipette so that the cut end is set in the water. Do this while holding the sprig and the pipette under water to avoid getting air bubbles in the pipette.

9. For each of the large test tube setups, insert an inverted pipette with an *Elodea* sprig into the sodium bicarbonate solution. Clamp the pipette to the ring stand so that the leaves of the sprig are hanging down from the pipette and completely immersed in the sodium bicarbonate. *Be sure to keep the open end of the pipette and the Elodea leaves below the level of sodium bicarbonate.*

10. Place one of the unwrapped setups in the dark as a negative control for photosynthesis.

11. For each of the remaining setups, position a light source containing a 200-watt bulb such that it faces the test tubes. Set a container of cool water between the tubes and the light source. *What is the purpose of the cool water container used in this system?*

12. After 45 minutes, determine the total amount of oxygen given off by each plant by measuring the displacement of water from the pipette in mL. Record your measurements in Table 12.2 below.

13. Using the results in Table 12.2, graph the action spectrum for photosynthesis in *Elodea* leaves in Figure 12.5. Place *Wavelength* on x-axis and *Rate of Photosynthesis* on the y-axis.

FIGURE 12.4 Set-up for the action spectrum.

Results

TABLE 12.2   Rate of photosynthesis using different color filters.

Color of light	Wavelength	Volume (mL) of oxygen released in 45 minutes = A	A/45 = Rate of photosynthesis (mL/min)
Violet			
Blue			
Green			
Yellow			
Red			
No color (set in light)			
No color (set in dark)			

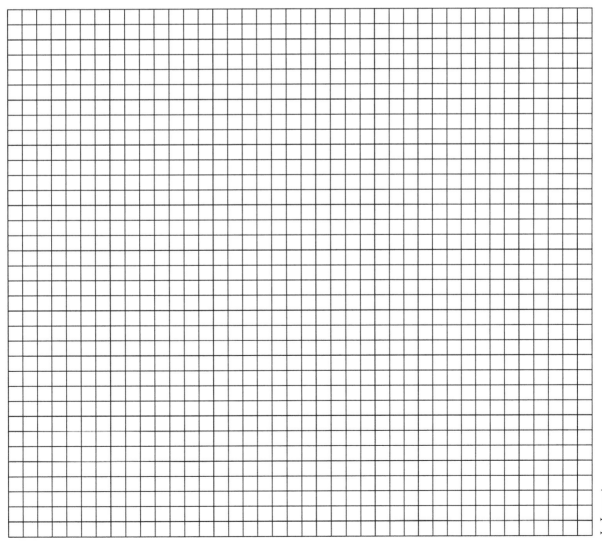

L. Jayant

FIGURE 12.5   Action spectrum of photosynthesis in *Elodea*.

Procedure 2b: The Absorption Spectrum

1. Fill cuvette 1 with water. This will be the **blank** sample.
2. Blend 10–15 fresh spinach leaves with 200 mL of water in a blender and filter the extract with a cheesecloth.
3. Dilute the spinach extract by adding about 9 mL of water to 1 mL of extract. The color should be pale green.
4. Fill cuvette 2 with the diluted extract. This will be the **test** sample.
5. Wipe the sides of the cuvettes with a Kimwipe.
6. Turn on the spectrophotometer and allow it to equilibrate for 15 minutes.
7. Set the wavelength at 400 nm.
8. Set the MODE to "absorbance."
9. Insert cuvette 1 (the water **blank** sample) into the sample holder. Set the absorbance at 0% to calibrate the spectrophotometer.

   What is the purpose of having one cuvette filled with only water?

10. Remove cuvette 1 and insert cuvette 2 (the **test** sample). Close the cover and record the absorbance reading in Table 12.3 below. Remove cuvette 2. Note: if this absorbance reading is above 0.7, dilute the spinach extract until this initial reading is equal to or below 0.7.
11. Increase the wavelength by 25 nm.
12. Repeat step 9 to re-calibrate the spectrophotometer.
13. Repeat step 10 and record the new absorbance in Table 12.3.
14. Continue to repeat steps 9 through 11, each time increasing the wavelength in 25 nm increments and recording the absorbance. Measure absorbance up to 725 nm. *Be sure to re-calibrate the spectrophotometer each time you change the wavelength.*
15. Using the results in Table 12.3, graph the absorption spectrum of the photosynthetic pigments obtained from spinach leaves in Figure 12.6. Place *Wavelength* on the x-axis and *Absorbance* on the y-axis.
16. Use the Internet to research the color that is correlated with each wavelength.

Results

TABLE 12.3   Absorbance of photosynthetic pigments in spinach extract.

Wavelength (nm)	Absorbance
400	
425	
450	
475	
500	
525	
550	
575	
600	
625	
650	
675	
700	
725	

Discussion

Discuss your results with members of your group. Does the absorption spectrum match the action spectrum? Explain your observations in the space provided.

_____

_____

_____

_____

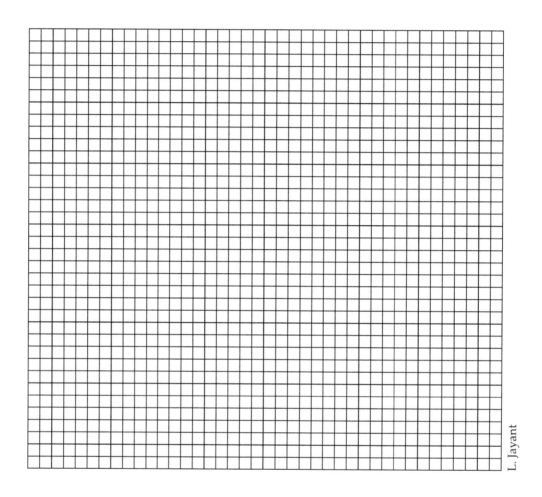

L. Jayant

FIGURE 12.6   Absorption spectrum of photosynthetic pigments in spinach leaves.

III. COMPREHENSION CHECK

1. Write a summary equation for photosynthesis.

2. Three pigment solutions are given to you. The first pigment, Ap, contains four –OH and two –NH$_2$ functional groups. The second pigment, Bp, has two –OH groups. The third pigment, Cp, has a very long hydrocarbon chain and is totally non-polar. You are told to separate these pigments by paper chromatography using a non-polar solvent. Predict the locations of the three pigments on the paper with respect to the solvent front after they separate. Explain your answer.

3. In response to shortened daylight and cool temperatures during the fall, tree leaves display a beautiful array of colors from red to orange to gold to yellow. Explain this phenomenon using your knowledge of photosynthesis and photosynthetic pigments.

4. Deep ocean dwelling red algae range in color from pink to red to dark purple. Identify the photosynthetic pigment that gives the red color to these algae. What colors and wavelengths of light do you think this pigment absorbs?

5. A new photosynthetic plant is discovered on Mars. It is blue in color. Draw the absorption spectrum and the action spectrum for the pigments in this plant.

CELLULAR RESPIRATION AND FERMENTATION

After completing this exercise, students should be able to do the following:

1. Distinguish between anaerobic and aerobic cellular respiration.
2. Summarize the reactions involved in generating ATP in the presence and absence of oxygen.
3. Distinguish between obligate aerobes, obligate anaerobes, and facultative anaerobes.
4. Name the stages of aerobic and anaerobic respiration and state the ATP yield in each.
5. Describe what happens in oxidation-reduction reactions.

I. INTRODUCTION

Whether prokaryotic or eukaryotic, unicellular or multicellular, all living organisms require energy to carry out the multitude of activities that enable them to survive. To meet their energy needs, the cells of an organism undergo some form of **cellular respiration**, a process that provides cells with usable energy on demand. Cellular respiration is an **exergonic** reaction that extracts chemical energy stored in the covalent bonds of organic nutrients such as sugars, fats, and amino acids, and transfers it into the bonds of the adenosine triphosphate molecule, or **ATP** (Figure 13.1). ATP provides the immediate energy that drives nearly all biochemical reactions in cells, including the synthesis of biological macromolecules, the active transport of substances across a plasma membrane, and the movement of cellular components.

The main fuel for the process of cellular respiration is glucose. Although a molecule of glucose contains a large amount of potential energy in its covalent bonds, it is difficult for a cell to obtain this energy when needed. Instead, glucose is broken down, or **catabolized**, through a series of metabolic reactions in which energy is slowly released in a controlled manner and ultimately transferred to ATP, a far more manageable molecule from which the cell can obtain the energy it needs to carry out its work.

Cellular respiration can take place either in the presence or in the absence of oxygen. **Aerobic cellular respiration** is the complete breakdown of glucose in the presence of oxygen to produce large amounts of ATP. Figure 13.2 depicts a summary equation for aerobic cellular respiration, although the actual process involves a number of complex metabolic reactions. Throughout the process, high-energy

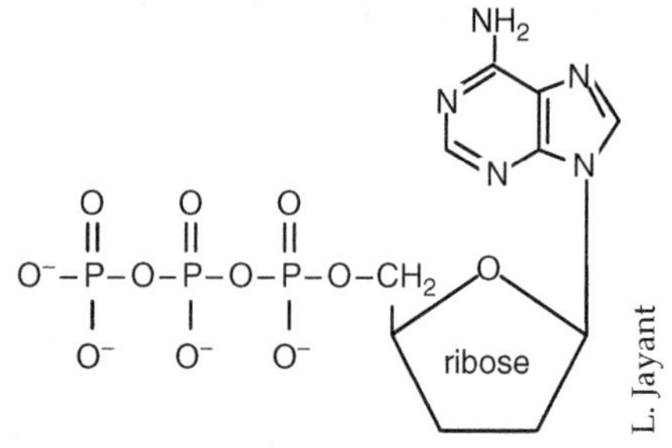

FIGURE 13.1 Structural formula for ATP.

FIGURE 13.2 Summary reaction for aerobic cellular respiration.

electrons are continually transferred in a series of chemical reactions known as **oxidation-reduction** reactions. When one substance loses electrons (*oxidation*), another substance must gain the electrons (*reduction*). During aerobic cellular respiration, glucose is completely oxidized to produce carbon dioxide. The final electron acceptor is oxygen, which is reduced to form water. As a result of this process, usable energy for the cell is harnessed in the phosphate bonds of ATP. For every molecule of glucose consumed during aerobic respiration, 36 to 38 molecules of ATP can be generated.

The first stage of cellular respiration is **glycolysis**, an anaerobic pathway in which one molecule of glucose is split into two molecules of **pyruvate** in a series of enzymatically catalyzed reactions. Glycolysis takes place in the cytosol and yields 2 ATPs per molecule of glucose (Figure 13.3). When oxygen is present, pyruvate (*or pyruvic acid*) proceeds to the **citric acid cycle**, another series of enzymatic reactions that take place in either the mitochondrial **matrix** of eukaryotic cells or the cytosol of prokaryotic cells. The citric acid cycle yields an additional 2 ATPs for every glucose molecule metabolized and generates most of the CO_2 produced. As bonds are broken between successive carbon atoms, high-energy electrons are transferred to coenzymes **NAD⁺** and **FAD, electron carrier** molecules that respectively become reduced to NADH and $FADH_2$. The reduced coenzymes transfer electrons to an **electron transport chain** on the inner membrane of the eukaryotic mitochondrion or on the inner surface of the prokaryotic plasma membrane. During electron transport, electrons are transferred to progressively lower energy levels through another series of oxidation-reduction reactions. Electron transport is coupled to the **chemiosmosis** of hydrogen ions in a process called **oxidative phosphorylation**, during which oxygen is reduced to form water, and energy released from the electron transport chain is used to generate ATP from ADP and inorganic phosphate. For every glucose molecule that is completely broken down, oxidative phosphorylation yields 32 to 34 molecules of ATP. Note that the major product of cellular respiration is ATP; water and carbon dioxide are by-products.

Most organisms require oxygen to meet their metabolic needs and are known as **obligate aerobes.** Alternatively, some organisms cannot survive in the presence of oxygen. These **obligate anaerobes** produce ATP exclusively through **anaerobic respiration**, a process that generates ATP in the absence of oxygen. One such organism is *Clostridium botulinum*, a bacterium that causes food poisoning.

At times when oxygen is lacking, some aerobic cells are able to generate usable energy by reverting to **fermentation**, an **anaerobic** process in which small amounts of ATP can be generated by the conversion of pyruvate to either lactic acid **(lactic acid fermentation)** or to ethanol and CO_2 **(alcoholic fermentation)** (Figure 13.4). For each glucose molecule used, fermentation yields two molecules

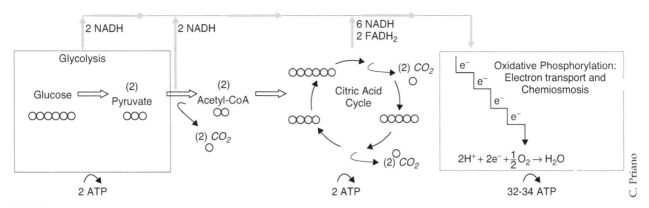

FIGURE 13.3   Overview of aerobic cellular respiration. → w/ oxygen

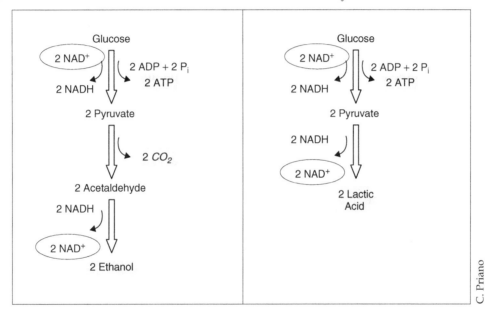

FIGURE 13.4   Alcoholic fermentation (left); lactic acid fermentation (right).

of ATP. Lactic acid fermentation occurs in some bacteria and in our muscle cells when ATP usage exceeds oxygen availability. **Facultative anaerobes**, such as yeast, can undergo aerobic respiration in the presence of oxygen, and alcoholic fermentation in the absence of oxygen.

II. LABORATORY ACTIVITIES

ACTIVITY 1: FERMENTATION IN YEAST USING FRUIT JUICE

In this experiment, you will measure carbon dioxide production when yeast is incubated with a common juice. Each lab group will test a different juice as a source of fuel in yeast.

Materials

water bath set at 42°C   7 grams active dry yeast
thermometer   600 mL beaker (to prepare yeast)
fermentation tubes (2 per group)   50 mL beakers (2)
distilled water   25 mL graduated cylinder
juice (different one for each group)   glass stirrer
pipettes   pH paper

Procedure

1. Mix 7 grams of yeast in 200 mL warm distilled water. Note the distinctive smell of the yeast suspension. Keep the yeast suspension at 42°C until use.

2. Pre-warm all juices to 42°C. Determine the number of grams of sugar in one milliliter of juice (obtain this information from the nutrition label on the juice container). Grams of sugar per mL of juice = _____.

3. Using the formula below, calculate the volume of juice (in milliliters) that contains 2 grams of sugar. Show your calculation:

$$\text{Formula 1:} \quad \frac{\text{grams of sugar}}{1 \text{ mL}} = \frac{2 \text{ grams}}{X \text{ ml}}$$

4. Label two 50 mL beakers 1 and 2. Label two fermentation tubes 1 and 2.

5. To beaker 1, add 18 mL of distilled water.

6. To beaker 2, add the volume of juice that contains 2 g of sugar, plus enough distilled water to bring the total volume to 18 mL. Indicate your juice sample in the second row of Table 13.1.

7. Determine the pH of each juice and record in Table 13.1. Note the smell of the juice.

8. To each beaker, add 2 mL of the yeast suspension from step 1 and stir. Note the smell.

9. Pour each 20 mL mixture into separate fermentation tubes (1 and 2).

10. Invert each tube to remove the air bubble from the vertical arm. The vertical arm should be completely filled with the mixture.

11. Set each fermentation tube in a water bath set to 42°C. *Answer the questions below.*

12. After 1 hour, measure the displacement in cm of liquid from the vertical arm of each fermentation tube. Record these measurements in Table 13.1. Note the smell of the mixtures.

13. Record the measurements that other lab groups obtain for their juices.

14. For each juice, calculate the volume of gas produced by using the formula for finding the volume of a cylinder:

$$\text{Formula 2:} \quad V = \pi r^2 h$$

where π is 3.14, r is the radius of the vertical arm of the fermentation tube in cm, and h is the height of displacement of liquid in cm. Record the volume of gas produced in Table 13.1.

15. In Figure 13.5, plot your results as a bar graph to show how different juices support yeast cell respiration. (Properly label your axes and include a title.)

What is the purpose of the experiment?_____

_____

What is your hypothesis?_____

_____

What is the independent variable?_____

What is the dependent variable?_____

What are the controlled variables?_____

_____

Results

TABLE 13.1   Volume of carbon dioxide gas produced by yeast.

Name of Juice	pH	Displacement of liquid (cm)	Volume of CO$_2$ produced (cm^3)
Control (water)			

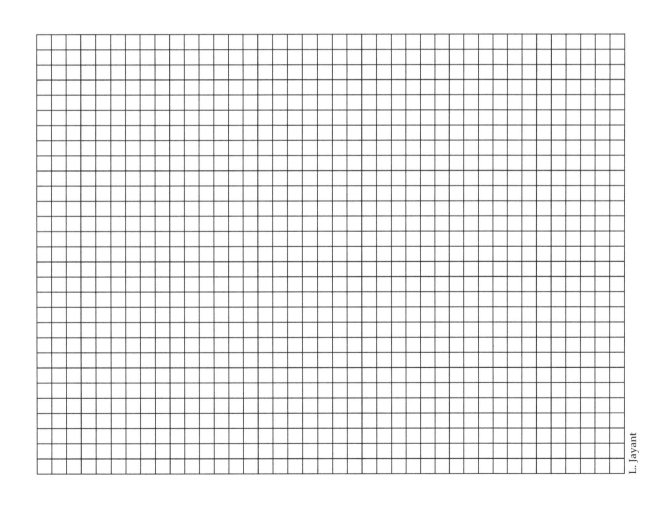

FIGURE 13.5   CO$_2$ production in yeast using different juices.

Discussion

1. What percent sugar solution did you use for your experiment?

2. Describe the smell of the juice/yeast mixture before and after the experiment.

3. What does displacement of liquid in the arm of the fermentation tube indicate?

4. Did pH have any effect on the outcome? Explain.

5. Do your results support your hypothesis? Explain.

6. Discuss the importance of this experiment.

ACTIVITY 2: OXIDATION OF APPLE SLICES

Oxidation occurs when one substance draws electrons from another. The substance that loses electrons becomes **oxidized**, while the one that gains the electrons becomes **reduced**. Aerobic respiration is an oxidation-reduction process in which oxygen is the **oxidizing agent**. Oxygen causes glucose to become oxidized, while the oxygen itself is reduced. Common oxidation reactions in which oxygen acts as an oxidizing agent include the rusting of iron and the formation of copper carbonate, the green coating that develops on copper surfaces (such as on the Statue of Liberty). In living cells, oxidation can often lead to the accumulation of **free radicals**, charged particles that can cause damage to the cells. **Antioxidants** are substances that prevent the buildup of harmful by-products of oxidation. Common antioxidants are citric acid and beta-carotene, found in many fruits and vegetables.

The discoloration of fruits is the result of oxidation. When the skin of a fruit is broken, and the plant tissue on the inside is exposed to oxygen, some phenolic compounds become oxidized, forming a brown pigment. In this experiment, you will observe the process of oxidation in apple slices that are coated with different substances, and examine which of these substances have antioxidant properties.

What is the purpose of the experiment?_____

_____

What is your hypothesis?_____

_____

Materials

lemon juice   knife
orange juice   distilled water
grape juice   one fresh apple (Macintosh or Red Delicious)
paper towel   **Do not use apples that have been developed to resist browning*

Procedure

1. Cut four slices of apple, approximately the same size and thickness.
2. Dip each slice in distilled water, lemon juice, orange juice, or grape juice. Remove the slice.
3. Set each slice side by side on a paper towel.
4. Observe the color changes in each apple slice every 5 minutes for 30 minutes. Use the chart in Figure 13.6 as a guide to rank the intensity of darkness as the color changes.
5. Record your observations in Table 13.2.

Results

TABLE 13.2   Intensity of discoloration in apple slices.

Sample	0 min	5 min	10 min	15 min	20 min	25 min	30 min
distilled water							
lemon juice							
orange juice							
grape juice							

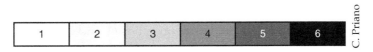

FIGURE 13.6   Color intensity scale for oxidation in apple slices.

Discussion

1. Which apple slices discolored at a faster rate?

2. Did any of the juices demonstrate antioxidant activity? Explain.

3. Do your results support your hypothesis? Explain.

4. Discuss the importance of this experiment.

ACTIVITY TO REINFORCE CONCEPTS

ACTIVITY 3: EFFECT OF TEMPERATURE ON FERMENTATION RATE IN YEAST

Materials

water bath set at 42°C	600 mL beaker
thermometer	50 mL beakers (2)
fermentation tubes (2 per group)	pipettes
distilled water	25 mL graduated cylinder
sucrose	glass stirrers
yeast suspension from Activity 1	ice bath

Procedure

1. Label two 50 mL beakers 1 and 2. Label two fermentation tubes 1 and 2.
2. To beaker 1, add 18 mL of distilled water.
3. To beaker 2, dissolve 2 g of sucrose in a total volume of 18 mL.
4. To each beaker, add 2 mL of the yeast suspension from Activity 1 and stir.
5. Pour each 20 mL sample into separate fermentation tubes (1 and 2).
6. Invert each tube to remove the air bubble from the vertical arm. The vertical arm should be completely filled with the mixture.

7. Each group will place its two fermentation tubes either in a water bath set at 42 °C; at room temperature; or in a 4 °C ice bath (use a 600 mL beaker containing ice and water). *Answer the questions below.*

8. After 1 hour, measure the displacement of liquid (in cm) in the vertical arm of each fermentation tube. Record these measurements in Table 13.3. Record the measurements of your classmates.

9. Using Formula 2 (see Activity 1), calculate the volume of gas produced in each tube. Record the volume of gas produced in Table 13.3.

10. Plot your results on a bar graph (Figure 13.7) to show how temperature affects yeast cell respiration. (Properly label your axes and include a title.)

What is the purpose of the experiment?_____

_____

What is your hypothesis?_____

_____

What is the independent variable?_____

What is the dependent variable?_____

What are the controlled variables?_____

_____

Results

TABLE 13.3 Volume of carbon dioxide gas produced by yeast.

Temperature	Displacement of liquid (cm)	Volume of CO_2 produced (cm^3)
4 °C		
25 °C		
42 °C		

Discussion

1. How did temperature affect the fermentation process? Explain.

2. Do your results support your hypotheses? Explain.

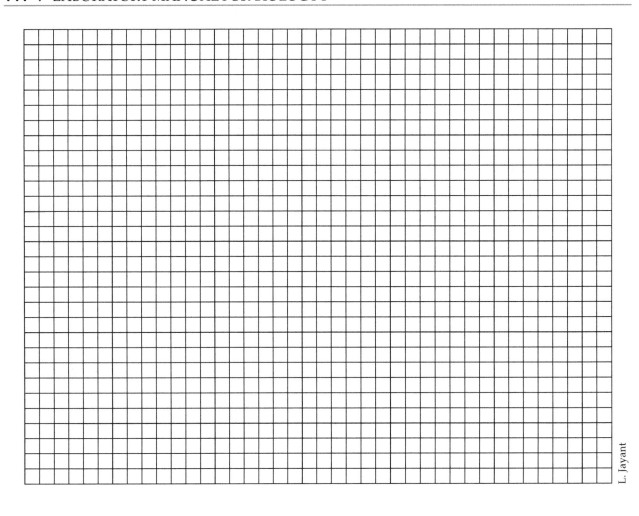

FIGURE 13.7 Affect of temperature on cellular respiration in yeast.

III. COMPREHENSION CHECK

Answer the following questions in complete sentences:

1. With respect to cellular respiration, what kind of an organism is yeast? Explain.

2. Why can't obligate aerobes survive using fermentation as a means of energy production?

3. What can you conclude from the gas displacement and from the smell of the juice in your experiment from Activity 1?

4. What is the literal meaning of the word *glycolysis*?

5. Describe the stages of aerobic cellular respiration and state the ATP yield in each.

6. Describe the difference between *oxidation* and *reduction* reactions.

7. Why are antioxidants good for your body?

8. During vigorous exercise, our muscles undergo fermentation and produce lactic acid. Why don't they produce alcohol?

CELLULAR DIVISION

After completing these laboratory exercises students should be able to:

1. Recognize the phases of the cell cycle and cellular division.
2. Identify the different stages of mitosis and meiosis.
3. Comprehend the events of mitosis and meiosis.
4. Distinguish between the stages of mitosis and of meiosis.

I. INTRODUCTION

Cell division plays a major role in the life of a cell. It is necessary for growth, development, and repair, and it is vital for reproduction. All organisms, whether prokaryotic or eukaryotic, must undergo some form of cellular division in order to sustain the organism and to propagate the species. This laboratory will focus on cell division in eukaryotic cells.

Cell division is an integral part of the **cell cycle,** a series of events in which a newly formed cell will again divide into two new cells. Figure 14.1 illustrates the eukaryotic cell cycle. A new cell that emerges in the cell cycle progresses through a number of stages until it divides again. The stages of the eukaryotic cell cycle are **G1, S, G2, mitosis,** and **cytokinesis**. During G1, a newly born cell grows in size and performs its normal functions. As the cell enters the S phase, the nuclear DNA containing the hereditary information of the cell begins to replicate by a process called DNA synthesis. By the end of the S phase, all DNA contained on **chromosomes** is completely duplicated, but not completely separated. Each half of a duplicated chromosome is called a **sister chromatid**. Sister chromatids are held together at a region called the **centromere**. During G2 of the cell cycle, the cell prepares to divide.

A. Mitosis

Mitosis is the division of the nucleus in a eukaryotic cell and is associated with growth, development, and repair of lost or damaged tissue. It is also a means of eukaryotic asexual reproduction. Mitosis involves the separation of duplicated chromosomes in a **parent cell** into two new **daughter cells** through the mitotic stages of **prophase, prometaphase, metaphase, anaphase,** and **telophase**.

147

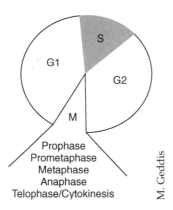

FIGURE 14.1 Cell cycle.

Cytokinesis, or division of the cytoplasm, occurs during telophase after the nucleus has divided. During mitosis, a series of events occurs in which the duplicated chromosomes of the parent cell are manipulated so that sister chromatids formed during interphase move apart. As a result, two genetically identical daughter cells emerge, each with a full complement of DNA identical to that of the parent cell.

B. Meiosis

Meiosis is a specific form of cell division that only occurs in specialized cells of sexually reproducing organisms. Unlike mitosis, meiosis is associated solely with gamete production and involves two separate cell divisions instead of one. All sexually reproducing organisms have two sets of chromosomes in each of their **somatic** cells, one set from each of two parents. As such, each chromosome of one set has a counterpart in the second set. A matched pair of chromosomes is referred to as a **homologous** pair. A cell that has two complete sets of homologous chromosomes is **diploid** *(has the diploid number of chromosomes)*. A **gamete**, the product of meiosis, has only one set of chromosomes and is **haploid** *(has the haploid number of chromosomes)*. The events of meiosis specifically work to separate homologous pairs of chromosomes in preparation for sexual reproduction.

As with all forms of cell division, cellular DNA must first be duplicated prior to meiosis. Cells that are destined to become gametes will enter interphase (G1, S, and G2) of the cell cycle, but will not enter mitosis. Instead, two successive meiotic cell divisions will take place. The first division, **meiosis I**, results in the separation of homologous pairs of duplicated chromosomes into two daughter cells. Because homologous pairs of chromosomes separate, the two resulting daughter cells are haploid, but still contain duplicated chromosomes (each with two sister chromatids). Because the chromosome number has been reduced to half, meiosis I is referred to as **reduction division**.

Each haploid daughter cell generated in meiosis I must undergo a second division, **meiosis II**, during which the sister chromatids become separated. The events of meiosis ultimately result in four genetically distinct haploid gametes, each with half the complement of DNA compared with that of the parent cell. *Note that there is no interphase between meiosis I and meiosis II.*

The individual stages of meiosis are similar to those of mitosis. Meiosis I includes **prophase I, prometaphase I, metaphase I, anaphase I,** and **telophase I** with **cytokinesis**. Meiosis II includes **prophase II, prometaphase II, metaphase II, anaphase II,** and **telophase II** with a second **cytokinesis**. Although corresponding stages of mitosis, meiosis I, and meiosis II entail general events that are similar, it is important to distinguish these stages from one another. This is especially true of meiosis I, during which homologous chromosomes first pair, then segregate. Pairing of two homologous chromosomes, or **synapsis**, occurs during prophase I to form a **tetrad**, a grouping of four sister chromatids. In the tetrad, corresponding segments from two homologous non-sister chromatids are aligned. In this conformation, an event known as **crossing over** can easily occur in which DNA from one chromatid can be exchanged with a matching segment from a homologous chromatid in the tetrad.

Crossing over between non-sister chromatids results in a mixing of hereditary material and genetic variation in the gametes that are produced. During anaphase I, homologous pairs of chromosomes segregate independently of all other pairs. This **independent assortment** of homologous chromosome pairs leads to a great deal of variety in the genetic makeup of gametes.

II. LABORATORY ACTIVITIES
ACTIVITY 1: IDENTIFY THE EVENTS OF MITOSIS
Materials

Figure 14.2
Figure 14.3

Procedure

1. Identify the distinguishing features of each stage of mitosis in the animal cell (Figure 14.2) and in the plant cell (Figure 14.3) below.
2. Complete Table 14.1.

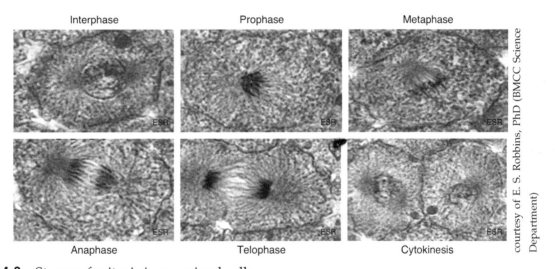

FIGURE 14.2   Stages of mitosis in an animal cells.

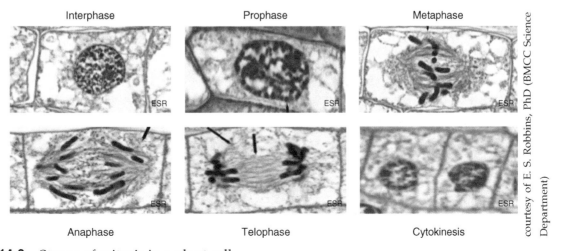

FIGURE 14.3   Stages of mitosis in a plant cell.

Results

TABLE 14.1   Stages of mitosis.

Label	Stage	Events
A	Prophase	
B	Prometaphase	Chromosomes begin to move.
C	Metaphase	
D	Anaphase	
E	Telophase	

ACTIVITY 2: IDENTIFY THE EVENTS OF MEIOSIS

Materials

Figure 14.4

Procedure

1. Identify the distinguishing features of each stage of meiosis below (Figure 14.4).
2. Complete Table 14.2.

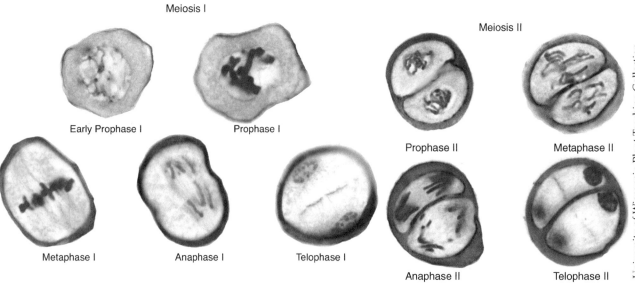

FIGURE 14.4   Stages of meiosis.

Results

TABLE 14.2   Stages of meiosis.

Label	Stage	Events
F	Prophase I	
G	Metaphase I	
H	Anaphase I	
J	Telophase I	
K	Prophase II	
L	Metaphase II	
M	Anaphase II	
N	Telophase II	

ACTIVITY 3: IDENTIFY THE STAGES OF MITOSIS AND MEIOSIS

Materials

Figure 14.5
Labels from Tables 14.1 and 14.2

Procedure

Use the labels and events in Tables 14.1 and 14.2 to identify different stages of mitosis in Figure 14.5a and meiosis in Figure 14.5b.

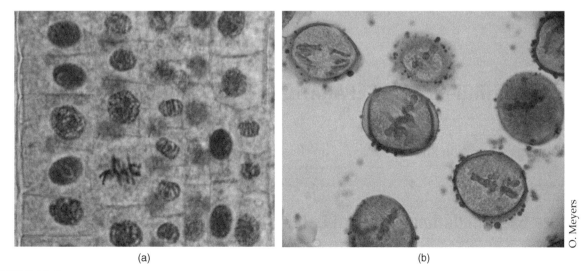

(a)   (b)

FIGURE 14.5   Images of cells in mitosis (a) and meiosis (b).

ACTIVITY 4: SLIDE PREPARATION OF ONION ROOT TIP SQUASH FOR IDENTIFICATION OF CELLS IN MITOSIS

Materials

water bath set at 60°C	double-welled plastic trays or Petri dishes	Feulgen stain
light microscope	forceps	1M HCl
prepared slides	razor blade	water
microscope slide	metric ruler	45% acetic acid
cover slip	1 mL pipette	
onion with root tip		
microtube		

Procedure

1. Using a razor blade, remove 1–2 mm of a root tip from an onion and place in a microtube. Note the orientation of the root tip (which region is closer to the onion versus the tip).
2. Place 1 mL of 1 M HCl in the microtube.
3. Place the microtube in a 60°C water bath for 10 minutes.
4. Transfer the root tip from the tube to a double-welled tray (or Petri dish) using forceps.
5. Cover the root tip with water and soak for 5 minutes.
6. Remove the tip from the water and transfer to another well. Cover the root tip with Feulgen stain for 15 minutes or until the end of the tip turns dark pink.
7. Rinse the root tip with water to remove excess stain.
8. Transfer the root tip to a microscope slide. With a razor blade, separate the dark pink end of the tip from the remainder of the root. Use only the end of the tip to make your squash.
9. Add 2 drops of 45% acetic acid to the tip end on the slide. Use a razor to cut the tip into several pieces.
10. Place a cover slip over the cut root tip and carefully press straight down with a pencil eraser until the pieces are spread out flat. Do not crack the cover slip or move it from side to side.
11. Observe the stained root tip under low and high power of the light microscope and record your results below in Figure 14.6.
12. Compare sections that were closer or further from the onion, as well as other students' preparations and the prepared slides of onion root tips.

Results

ACTIVITY 5: CHROMOSOME SEPARATION SIMULATION

Materials

chromosome separation kit	holders for beads
red beads	white connectors
yellow beads	small magnetic stir bars

Procedure 5a. Simulation of Animal Cell Mitosis

1. Obtain one chromosome separation kit to simulate mitosis.
2. As you simulate each phase of mitosis, record your observations by sketching each phase in Figure 14.7 below.
3. To simulate a eukaryotic cell, place a shorter length of string in a circle (to represent a nucleus) inside a longer circle of string (representing the plasma membrane).

Results

M. Geddis

FIGURE 14.6   Recorded observations of mitosis in stained onion root tip cells.

4. To simulate chromosomes, assemble beads (think of them as individual traits) along two holders. One strand of beads will have yellow "traits" and one will have red "traits."

5. Simulate the replication of these two chromosomes by first making an additional two identical strands. Then, connect the two red strands together with one white connector (the "centromere") and connect the two yellow strands together with a second white connector. Once the strands are connected, you should have two duplicated chromosomes (one red and one yellow). Each individual strand represents a sister chromatid of a duplicated chromosome.

6. Place the two duplicated chromosomes inside the inner circle (nucleus). As you simulate prophase, remove the inner circle to represent the fragmentation of the nuclear envelope.

7. Move the chromosomes around to represent prometaphase.

8. Line the two duplicated chromosomes along the center of the cell to represent metaphase.

9. To represent anaphase, pull the individual chromatids apart by removing the "centromere" connectors and begin moving each separated set of strands toward opposite borders of the cell. There are now four daughter chromosomes.

10. To represent telophase, move the daughter chromosomes all the way to opposite poles of the cell. Using two short pieces of string, arrange two new "nuclear envelopes" around the chromosomes at each pole.

11. To demonstrate cytokinesis, arrange the "plasma membrane" string to show a cleavage furrow, and pinch off each new cell. Note the appearance of the "daughter cells" compared with the original "parent cell."

12. Using one of your daughter cells, repeat steps 5–11. When completed, note the appearance of chromosomes in the new daughter cells after a second round of mitosis.

FIGURE 14.7   Recorded observations of mitosis simulation.

Procedure 5b. Simulation of Animal Cell Meiosis

1. As you simulate each phase of meiosis, record your observations by sketching each phase in Figure 14.8 below.

2. Use the same two chromosomes, one red and one yellow, that you constructed in step 4 of Procedure 5a. These will represent one homologous pair of chromosomes.

3. Simulate a eukaryotic cell membrane and a nuclear envelope as in step 3 of Procedure 5a above.

4. To simulate DNA replication, repeat step 5 of Procedure 5a above.

5. To simulate a tetrad, pair the two homologous chromosomes side by side and connect them.

6. To simulate crossing over, exchange two or three of the red beads from the end of a red chromatid with the corresponding yellow beads on a non-sister chromatid strand.

7. Place the tetrad inside the inner circle (nucleus). As you simulate prophase I, remove the inner circle to simulate the fragmentation of the nuclear envelope.

8. Move the tetrad around to simulate prometaphase I.

9. To simulate metaphase I, bring the tetrad to the center of the cell.

10. To simulate anaphase I, pull the two homologous chromosomes apart and begin moving each duplicated chromosome of the homologous pair toward opposite borders of the cell. *Do not separate the sister chromatids.*

11. To simulate telophase I, move the two homologous chromosomes all the way to opposite poles of the cell. Using two short pieces of string, arrange two new "nuclear envelopes" around the duplicated chromosomes at each pole.

12. To demonstrate cytokinesis I, arrange the "plasma membrane" string to show a cleavage furrow, and pinch off each new cell. Note the appearance of the "daughter cells."

13. For each of the two daughter cells, remove the inner circle to simulate prophase II.

14. To simulate metaphase II, move each duplicated chromosome to the center of each cell.

15. To simulate anaphase II, pull the individual chromatids apart by removing the "centromere" connector and begin moving each separated strand toward opposite borders of each cell.

16. To simulate telophase II, move the daughter chromosomes all the way to opposite poles of each cell. Use two short pieces of string to arrange "nuclear envelopes" around the chromosomes..

17. To demonstrate cytokinesis II, arrange the "plasma membrane" string of each cell to show a cleavage furrow. Pinch off the newly formed "daughter cells."

18. When completed, note the appearance of the four haploid "daughter cells" compared with the original "parent cell." Also note the appearance of daughter cells after meiosis I and meiosis II.

Results

FIGURE 14.8   Recorded observations of meiosis simulation.

Procedure 5c. Simulation of Microtubule Polymerization and Depolymerization

1. As you simulate spindle fiber movement, record your observations in Figure 14.9 and identify the stage of mitosis.

2. Place a long length of string in a circle to represent the plasma membrane of a cell.

3. Assemble a duplicated chromosome by connecting two strands of beads as in Procedure 5a. Place the duplicated chromosome inside the cell.

4. Place one magnetic stir bar at one pole of the cell and one at the opposite pole.

5. Continue to attach stir bars from each pole of the cell until a series of stir bars (the microtubule "spindle fiber") contacts the white connector "centromere" of the duplicated chromosome.

6. Continue to add or remove stir bars from each "spindle fiber" until they hold the chromosome at the center of the cell (these represent kineticore spindle fibers).

7. Construct two separate strings of stir bar "spindle fibers" that span directly from one pole to the other, without touching the chromosome (these represent non-kineticore spindle fibers).

8. On the duplicated chromosome, pull the white connector apart. As you move the sister chromatids toward opposite poles of the cell, start to remove stir bars from the kineticore end of the kineticore "spindle fibers" to simulate depolymerization of the microtubules.

9. As you depolymerize the kineticore microtubules, add more stir bars to the longer non-kineticore "spindle fibers." These will continue to grow longer, pushing on opposite poles of the cell to elongate the cell and initiate cytokinesis.

Results

FIGURE 14.9   Recorded observations of microtubule simulation.

TABLE 14.3 Compare mitosis and meiosis.

Property	Mitosis	Meiosis
DNA replication		
Divisions		
Synapsis and crossing over		
Daughter Cells, genetic composition		
Role in animal body		

ACTIVITY 6: COMPARE AND CONTRAST THE EVENTS OF MITOSIS AND MEIOSIS

In Table 14.3, list similarities and differences between mitosis from meiosis.

ACTIVITY TO REINFORCE CONCEPTS
ACTIVITY 7: MITOSIS AFTER MEIOSIS
Materials

dissecting microscope
depression slides
sea urchin gametes (if available)
pipette

video of sea urchin fertilization and
  early development

Procedure

1. Your instructor will show a video of sea urchin fertilization and early development. Observe the movement of the chromosomes.
2. Obtain male and female sea urchin gametes (products of meiosis).
3. Under the dissection microscope, place on drop of each gamete solution on a depression slide.
4. Observe for fertilization (formation of a polar body at about 5–7 minutes). Sketch your observations in Figure 14.10.
5. Observe for cell division (mitosis; first cleavage at about 50–70 minutes). Sketch your observations in Figure 14.10.

Results

M. Geddis

FIGURE 14.10 Recorded observations of sea urchin fertilization.

III. COMPREHENSION CHECK

1. What are the stages of the eukaryotic cell cycle?

2. What is the difference between mitosis and cytokinesis?

3. Why does mitosis occur?

4. What is the end product of mitosis?

5. Why does meiosis occur?

6. What is the end product of meiosis?

7. What is the difference between meiosis I and meiosis II?

8. What is the purpose of meiosis I? What is the purpose of meiosis II?

9. Explain the difference between metaphase of mitosis, metaphase of meiosis I, and metaphase of meiosis II.

10. Which events of meiosis lead to chromosomal variation in offspring?

11. Consider a eukaryotic cell that has only two chromosomes. Describe what happens during each stage of the cell cycle, including the individual stages of interphase, mitosis, and cytokinesis.

DNA ISOLATION AND MODELING

After completing these laboratory exercises students should be able to:
1. Recognize that all living organisms contain DNA.
2. Identify the components and structure of a DNA molecule.
3. Assemble a nucleotide and a double helical DNA molecule using paper models.
4. Compare genomic DNA with plasmid DNA.
5. Isolate genomic DNA from plant cells and plasmid DNA from bacteria.

I. INTRODUCTION

DNA, or deoxyribonucleic acid, is the heredity material found in all living organisms. Inherited traits are passed from parent to offspring *via* DNA. It contains the genetic blueprint that determines the development and functioning of an organism, whether a simple unicellular bacterium or a complex multicellular organism such as a human.

The overall chemical structure and properties of DNA are well conserved in all organisms, both prokaryotic and eukaryotic. All DNA is composed of two chains of polymerized **deoxyribonucleotide** monomers. A deoxyribonucleotide is made up of a five-carbon pentose sugar called **deoxyribose** (Figure 15.1), a nitrogen-containing **base**, and three bound phosphate groups known as a **triphosphate**. The five carbons of the deoxyribose are numbered C_1 through C_5, as shown in Figure 15.1 The nitrogen-containing base in a deoxyribonucleotide can be either a purine (adenine or guanine), or a pyrimidine (cytosine or thymine) (Figure 15.2).

Figure 15.3 shows the molecular structure of a deoxyribonucleotide monomer, the basic subunit of DNA. On the deoxyribose sugar of each nucleotide monomer, a nitrogen-containing base is covalently bound to the C_1 carbon atom. A triphosphate is covalently bound to the C_5 carbon atom, also called the 5′ carbon (*pronounced "5-prime carbon"*). This side of the nucleotide is referred to as the 5′ end. A hydroxyl (OH) group is covalently bound to the 3′ carbon at the 3′ end of the nucleotide (Figure 15.4). As a DNA molecule is polymerized, deoxyribonucleotide monomers are added so that the 5′ end of an incoming nucleotide forms a covalent bond with the 3′ end in the growing DNA. Thus, DNA always elongates in the 5′ to 3′ direction.

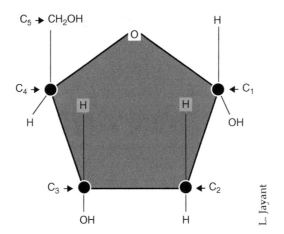

FIGURE 15.1   Deoxyribose.

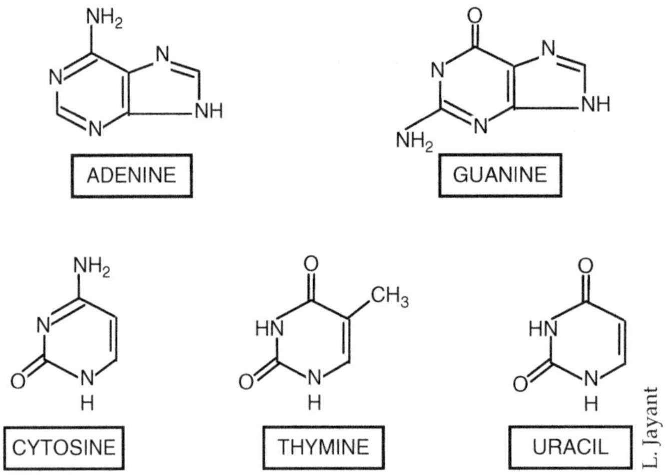

ADENINE

GUANINE

CYTOSINE

THYMINE

URACIL

FIGURE 15.2   Chemical formulas of nitrogen-containing bases.

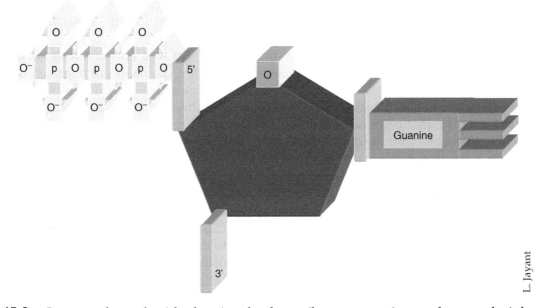

FIGURE 15.3   Cartoon of a nucleotide showing the deoxyribose sugar, nitrogen base, and triphosphate.

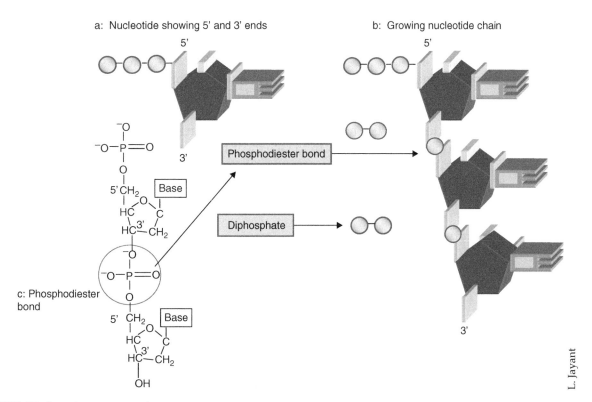

FIGURE 15.4 The DNA polymer.

Figure 15.4 illustrates the elongation of a DNA polymer. When one nucleotide binds with another nucleotide, a covalent bond is formed and a molecule of water is released *(dehydration synthesis)*. The resulting bond between the two nucleotides is called a **phosphodiester bond** (Figure 15.4c).

A DNA molecule is composed of two polynucleotide **strands** that are held together by hydrogen bonds and has the conformation of a **double helix** (Figure 15.5). The two strands are **antiparallel**, meaning they are oriented in opposite directions, and they are **complementary**, meaning that the

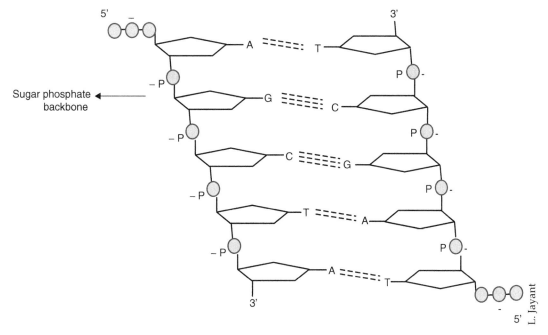

FIGURE 15.5 Double-stranded DNA molecule.

bases in the two strands pair up specifically. In opposing strands, adenine (A) pairs with thymine (T), and guanine (G) pairs with cytosine (C). The A:T base-pairs are held together by two hydrogen bonds, whereas the C:G base-pairs are held together by three hydrogen bonds. The basic structure of DNA can be divided into an external hydrophilic sugar-phosphate backbone and an internal hydrophobic region containing the paired nitrogen-containing bases. In the sugar-phosphate backbone, deoxyribose sugar and phosphate components alternate. Because the phosphate groups in the backbone are negatively charged, the exterior of the molecule is polar, rendering the DNA soluble in water. The bases, however, are hydrophobic, causing them to face the interior of the double helix and facilitating base pairing between the two complementary strands.

II. LABORATORY ACTIVITIES

ACTIVITY 1: SIMULATING DNA

Materials

cutouts at the end of this exercise   glue
scissors   colored pencils

Procedure 1a: Construction of Simulated Deoxyribonucleotides

1. Using the cutout pages at the end of this Laboratory Exercise, cut out one deoxyribose sugar, three phosphate molecules, and one nitrogen-containing base.
2. Number the carbons in the deoxyribose sugar, C_1 to C_5.
3. Glue the base onto the C_1 carbon of the deoxyribose such that the uneven end of the base is pointing away from the sugar.
4. Glue the three phosphates together in a row to make a triphosphate. Color the phosphates blue to indicate the negative charge.
5. Glue the triphosphate to the C_5 carbon of the deoxyribose.
6. Repeat steps 1 through 5 to construct the remaining deoxyribonucleotides. Use Figure 15.4 as a guide.

Procedure 1b: Construction of a Simulated DNA Model

1. Using six simulated deoxyribonucleotides from Procedure 1a, construct a polynucleotide chain. Refer to Figures 15.4 and 15.5. As a phosphodiester bond is formed between two nucleotides, remove two phosphates from the new nucleotide.
2. Construct a complementary polynucleotide chain that runs antiparallel to the first chain. Make sure that the bases are complementary in the two strands.

ACTIVITY 2: ISOLATION OF DNA FROM STRAWBERRIES AND BANANAS

Do fruits such as strawberries and bananas have DNA? DNA is present in all living organisms. Strawberries are well suited for DNA isolation because they are **polyploidy**. This means that they have more than two sets of chromosomes. DNA can be isolated from the cells of these fruits by breaking the plasma membranes and the nuclear membranes, removing the cellular debris, and forming an insoluble DNA precipitate that separates from an aqueous medium. Phospholipid membranes can be disrupted by a detergent such as soap or shampoo. Sodium chloride (NaCl) helps to remove proteins that are bound to the DNA and helps to separate proteins from DNA when mixed with alcohol.

Materials

ice bath	strawberries
blender	bananas
600 mL beaker	ice-cold isopropanol
100 mL graduated cylinder	NaCl
test tubes and rack	distilled water
dishwashing soap	beakers for mixing solutions
cheesecloth	glass rod

Procedure

1. **CAUTION: DO NOT EAT OR TASTE THE FRUIT USED IN THIS LABORATORY.**
2. To break up the fruit into individual cells and to break the cell walls, use a blender to blend 5 strawberries, 1 banana, and about 200 mL of water. The entire class will use this fruit mixture.
3. Pour the blended mixture into a 600 mL beaker.
4. Prepare 100 mL of a 3% NaCl solution.
5. To 50 mL of the NaCl solution, add 10 mL of concentrated dishwashing soap and 40 mL of distilled water. Mix gently with a glass rod until the detergent is uniformly mixed. Avoid frothing.
6. In a test tube, add 10 mL of the fruit mixture and 10 mL of the salt/soap mixture. Swirl the tube very gently until the solution is viscous.
7. Filter this mixture through cheesecloth and collect the filtrate in a separate test tube.
8. To 5 mL of filtrate, gently add 10 mL of ice-cold isopropanol by slowly dripping it down the side of the test tube. DO NOT MIX. The isopropanol should overlay the filtrate.
9. DNA will appear as a gelatinous precipitate at the junction between the two liquids (it will appear red here because of the color of the strawberries).
10. The DNA precipitate can be spooled on a glass rod.

ACTIVITY 3: ISOLATION OF PLASMID DNA FROM BACTERIA

Some bacteria contain extrachromosomal DNA called plasmids. Plasmids are circular DNAs that replicate independent of the chromosomal DNA and carry genes for special characters, such as resistance to antibiotics. Plasmids are useful tools in molecular biology. Plasmid DNA can be easily isolated from bacteria and further modified for use in recombinant DNA technology. Foreign genes can be inserted into plasmids, then amplified in bacterial hosts as the bacteria reproduce. Isolation of plasmid DNA is much simpler than isolation of intact chromosomal DNA. Plasmids are relatively small and the circular DNA has a supercoiled structure. These properties contribute to making plasmid DNA more resistant to breaking apart during isolation procedures.

Materials

LB/amp culture of bacteria transformed with pUC19 plasmid DNA (grown overnight)	microcentrifuge tubes
ice	microcentrifuge
800 mM NaOH	micropipettes (P1000 and P200)
4% SDS	micropipette tips
neutralization buffer (5 M potassium acetate, pH 5.5)	vortex
resuspension buffer: (100 μL solution A* + 10 μL RNase) [* solution A: 25mM Tris-HCl, pH 8.0]	ice cold isopropanol 70% ethanol

Procedure

Note: Maintain sterile conditions. Discard all used materials into a red biohazard safety bag.

1. *Your instructor will demonstrate the proper use of the micropipettes. For volumes between 200 μL and 1 mL, use the P1000. For volumes between 20 μL and 200 μL, use the P200.*
2. Pipette 1 mL of the overnight culture into a microcentrifuge tube using a P1000 micropipette.
3. Cap the tube and place it in a microcentrifuge, making sure that the instrument is balanced. *Your instructor will demonstrate how to balance and operate the microcentrifuge.*
4. After all tubes are balanced, secure the lid of the microcentrifuge and spin at 12,000 rpm for 1 minute. A bacterial pellet will form at the bottom of the tube.
5. Use the P1000 micropipette to aspirate the supernatant (fluid). Do not disturb the pellet at the bottom of the tube.
6. Add an additional 1 mL of the bacterial culture to the same tube and repeat steps 2 through 5.
7. Add 150 μL of resuspension buffer. Resuspend the pellet using a vortex (be sure to recap the tube). The bacteria should be completely resuspended. No clumps should be visible.
8. Prepare fresh lysis buffer by mixing 50 μL of 800 mM NaOH and 50 μL of 4% SDS.
9. Add 100 μL of freshly prepared lysis buffer from step 8 to the bacterial suspension. Mix **gently** by inverting the capped tube 5-6 times. The mixture should appear translucent and viscous.
10. Add 150 μL of neutralization buffer and mix **gently** by inverting 5-6 times. The mixture should contain a white precipitate of denatured proteins and cellular debris.
11. To remove the bacterial debris, spin in a microcentrifuge for 15 minutes at 12,000 rpm. Transfer the clear supernatant to a new microcentrifuge tube. Be sure not to pick up any debris. *Important: Your DNA is dissolved in the clear supernatant. Save the supernatant in the new tube and discard the old tube containing the cellular debris pellet.*
12. To the clear supernatant, add 400 μL of ice cold isopropanol. Mix thoroughly by inverting 10 times. Incubate at room temperature for 1 minute. The plasmid DNA is insoluble in the isopropanol and will precipitate out of solution.
13. Collect the precipitated DNA by spinning in a microcentrifuge for 20 minutes at the highest speed possible. The DNA pellet will not be visible, but will be lying against the lower back wall of the microcentrifuge tube. Carefully aspirate the isopropanol using a micropipette. Take care not to touch the pellet.
14. To rinse the DNA, add 500 μL ice cold 70% ethanol to the tube and mix by inverting 5-6 times.
15. Collect the DNA by spinning in a microcentrifuge for 10 minutes at the highest speed possible.
16. Carefully pour off the ethanol. Leaving the cap open, set the tube down for 10 minutes to allow the DNA to air dry.
17. Add 50 μL of distilled water to dissolve the DNA. Mix carefully and store the DNA in a capped microcentrifuge tube at 4°C for use in Laboratory Exercise 17.

III. COMPREHENSION CHECK

1. DNA is a polymer made of _____.

2. In a nucleic acid chain, the monomers are held together by _____ bonds.

3. DNA is made of two strands held together by _____.

4. Two DNA strands are *complementary* and *antiparallel*. Explain the meaning of these terms.

5. During DNA isolation, detergent is used to break the plasma membrane of the cells. What property of soap enables this process?

6. At the end of the DNA extraction, why was ice-cold 70% ethanol added after the isopropanol step?

7. Given below is a sequence of DNA. Complete the sequence of the complementary strand.
   5' - AGCCCTGACGTGCCAGT - 3'

8. What is the difference between plasmid DNA and the genomic DNA?

9. To isolate plasmid DNA, we add NaOH and SDS. Why don't we add NaOH to isolate genomic DNA? Explain.

10. Isolated DNA is a white gelatinous precipitate. If we put this precipitate under the microscope, can we see the double helix? Why or why not?

DEOXYRIBOSES

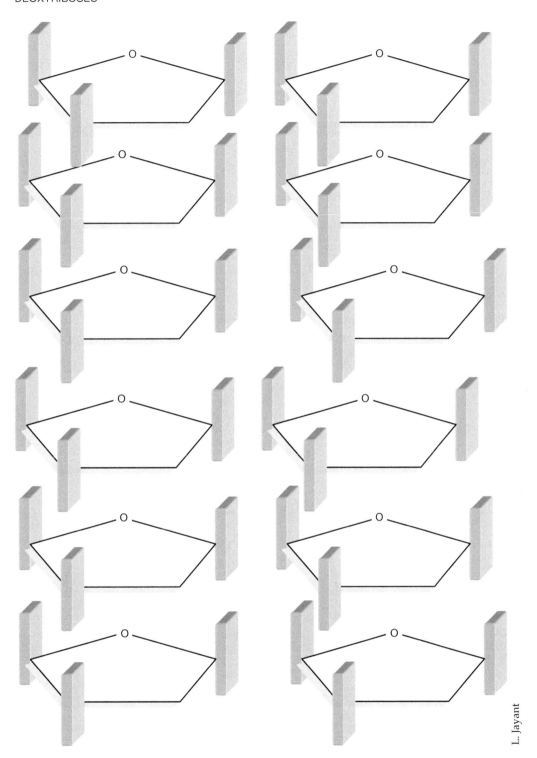

L. Jayant

BASES

PHOSPHATES

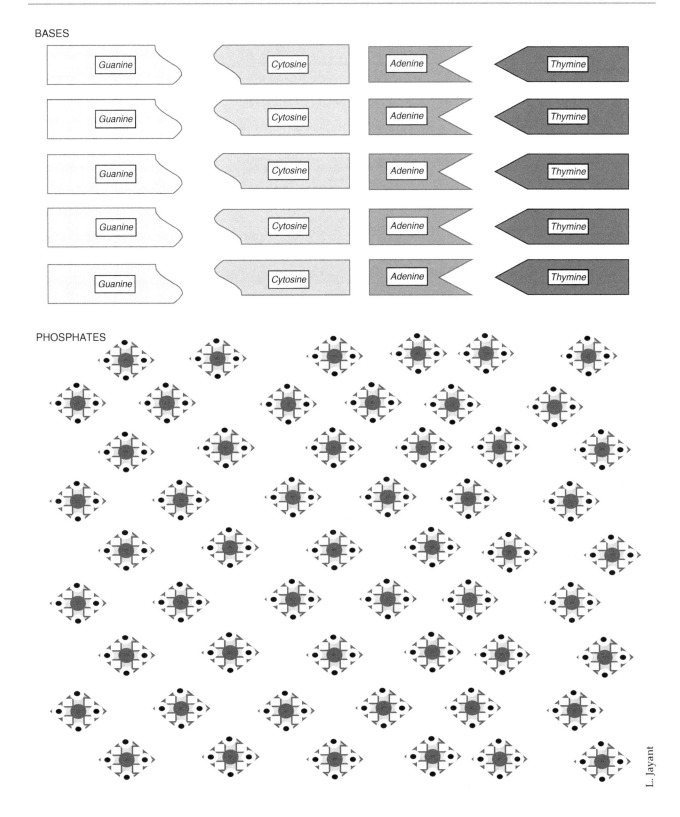

L. Jayant

TRANSCRIPTION, TRANSLATION, AND PROTEIN STRUCTURE

After completing this exercise, students should be able to do the following:

1. Describe how genetic information is transcribed from DNA to RNA.
2. Describe how the genetic code is translated from RNA into protein.
3. Distinguish between primary, secondary, tertiary, and quaternary protein structure.

I. INTRODUCTION

A **gene** can be defined in a number of ways: as a unit of inheritance, as a position or **locus** on a chromosome that is associated with a characteristic or trait, or as a segment of DNA that carries information, usually for making a protein. The molecular basis for the inheritance of your traits lies ultimately in the organization of DNA in your genes. The appearance of your traits depends on the proteins that are generated from active gene sequences.

Gene expression refers to the transfer of information from DNA in a gene to a product. In all living organisms, genetic information is processed the same way: from DNA to RNA to protein (Figure 16.1). Whereas DNA holds and preserves genetic instructions for making proteins, RNA carries out the instructions. Three types of RNA work to produce protein products dictated by a gene: A **messenger RNA** (mRNA) transfers the encoded information from the DNA to a ribosome, the site of protein synthesis; **ribosomal RNA** (rRNA) is part of the ribosome, and provides an anchor for the mRNA to attach; and **transfer RNA** (tRNA) delivers amino acids to the ribosome.

Transcription is the synthesis of RNA from a DNA template and is catalyzed by the enzyme **RNA polymerase**. When a gene is expressed, mRNA is transcribed from a template DNA strand in the gene. RNA nucleotides are incorporated into a growing mRNA according to the nucleotide sequence in the DNA template. RNA nucleotides A, C, G, and U pair up with DNA nucleotides T, G, C, and A respectively. Unlike DNA, RNA is synthesized as a single-stranded molecule. As a newly synthesized mRNA elongates, it is released from the DNA template in the gene. Because the mRNA is complementary to the DNA template, it carries the same information as the gene.

171

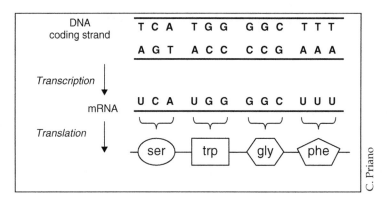

FIGURE 16.1 Flow of genetic information.

The **genetic code** (Figure 16.2) correlates the language of nucleic acids with the language of proteins. Each nucleotide triplet in the code, or **codon**, corresponds to one of 20 amino acids. Note that the code is **degenerate**, which means that there can be more than one codon for each amino acid. The codon AUG is an initiation or **start codon** that begins the synthesis of a protein. Three codons, UAG, UGA, and UAA are termination or **stop codons** that signal the end of an amino acid sequence in a growing polypeptide. Every gene contains a coding sequence that is composed of successive codons. During transcription, these codons are transferred from the DNA to mRNA. The mRNA in turn carries the genomic sequence of codons to the ribosome, where the encoded protein can be assembled in the process of **translation**. In prokaryotic cells, transcription generates mRNAs in which the codon sequences are uninterrupted. In eukaryotic cells, however, coding sequences, or **exons**, of a gene are often interrupted by non-coding sequences, or **introns**. These introns are excised from the functional mRNA prior to translation.

Translation of the genetic code from mRNA into protein takes place at the ribosome. Transfer RNAs (tRNAs) bridge the gap between the codons in the mRNA and the amino acids in the resulting protein. Each of the 20 amino acids has a specific tRNA to which it can attach. For every codon in the genetic code, there is a corresponding tRNA that contains a complementary nucleotide triplet called an **anticodon**. At the ribosome, the anticodon of a tRNA will pair only with its complementary codon in the mRNA, bringing the proper amino acid into a growing polypeptide chain.

	U	C	A	G	
U	UUU } Phe UUC UUA } Leu UUG	UCU UCC } Ser UCA UCG	UAU } Tyr UAC UAA - - Stop UAG - - Stop	UGU } Cys UGC UGA - - Stop UGG - - Trp	U C A G
C	CUU CUC } Leu CUA CUG	CCU CCC } Pro CCA CCG	CAU } His CAC CAA } Gln CAG	CGU CGC } Arg CGA CGG	U C A G
A	AUU AUC } Ile AUA AUG - - Met or Start	ACU ACC } Thr ACA ACG	AAU } Asn AAC AAA } Lys AAG	AGU } Ser AGC AGA } Arg AGG	U C A G
G	GUU GUC } Val GUA GUG	GCU GCC } Ala GCA GCG	GAU } Asp GAC GAA } Glu GAG	GGU GGC } Gly GGA GGG	U C A G

FIGURE 16.2 The genetic code.

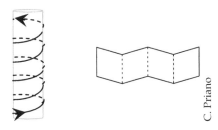

C. Priano

FIGURE 16.3   Left, alpha helix; right, pleated sheet.

Before a protein can be functional, it must be folded into a characteristic conformation that will enable it to perform its task. Proteins can adopt several levels of protein structure. The linear sequence of amino acids in a completed polypeptide is referred to as its **primary structure**. The **secondary structure** of a protein refers to a two-dimensional folding of the linear sequence of amino acids into either a coiled structure called an **alpha helix**, or a folded structure called a **pleated sheet** or beta sheet (Figure 16.3). Soluble proteins fold even further into **tertiary structure**, in which hydrophobic amino acids turn toward the core of a globular conformation. **Quaternary structure** refers to proteins that are composed of two or more polypeptide subunits. Different levels of protein structure are kept stable by various bonds, including hydrogen bonds, covalent disulfide bridges that form between certain amino acids, and ionic interactions.

Hemoglobin is an iron-containing protein found in red blood cells and functions in the transport and delivery of oxygen. Adult human hemoglobin (hemoglobin A) is a **tetramer** composed of four polypeptide subunit chains; two **alpha chains** and two **beta chains**. The alpha subunit of hemoglobin is 141 amino acids long and is encoded by either of two genes located very close together on the short arm of chromosome 16. These two genes comprise the alpha-globin **locus**. The beta subunit of hemoglobin is 146 amino acids long and is encoded by the beta-globin gene located on the short arm of chromosome 11 (Figure 16.4). Each of the four polypeptide subunits folds into a characteristic secondary structure made up of eight alpha helices. Each helical segment is connected by a short non-helical segment. Tertiary protein structure in each chain enables the incorporation of an iron containing **heme** group that can bind one molecule of molecular oxygen. The four subunits associate into a quaternary conformation. Each hemoglobin tetramer can carry four molecules of oxygen.

II. LABORATORY ACTIVITIES

ACTIVITY 1: TRANSCRIPTION AND TRANSLATION OF THE BETA-GLOBIN GENE

The entire DNA coding sequence of the beta-globin gene is shown in Figure 16.4 (introns have been eliminated). The initiator ATG codon and the termination TAA codons are included. In this activity, you will first **transcribe** the messenger RNA (mRNA) for the beta-globin peptide, then **translate** the mRNA into protein. You will then simulate the different levels of protein structure adopted by beta-globin.

Materials

DNA gene sequence for beta-globin	color-code for amino acids
genetic code	drinking straws
	scissors
pipe cleaners	beads (20 different colors)

Hemoglobin A (beta subunit) (146 amino acids; 438 coding nucleotides)[1]:

```
                                                             60
                                                             |
ATG GTG CAC CTG ACT CCT GAG GAG AAG TCT GCG GTT ACT GCC CTG TGG GGC AAG GTG AAC

                                                             120
                                                             |
GTG GAT GAA GTT GGT GGT GAG GCC CTG GGC AGG CTG CTG GTG GTC TAC CCT TGG ACC CAG

                                                             180
                                                             |
AGG TTC TTT GAG TCC TTT GGG GAT CTG TCC ACT CCT GAT GCA GTT ATG GGC AAC CCT AAG

                                                             240
                                                             |
GTG AAG GCT CAT GGC AAG AAA GTG CTC GGT GCC TTT AGT GAT GGC CTG GCT CAC CTG GAC

                                                             300
                                                             |
AAC CTC AAG GGC ACC TTT GCC ACA CTG AGT GAG CTG CAC TGT GAC AAG CTG CAC GTG GAT

                                                             360
                                                             |
CCT GAG AAC TTC AGG CTC CTG GGC AAC GTG CTG GTC TGT GTG CTG GCC CAT CAC TTT GGC

                                                             420
                                                             |
AAA GAA TTC ACC CCA CCA GTG CAG GCT GCC TAT CAG AAA GTG GTG GCT GGT GTG GCT AAT

GCC CTG GCC CAC AAG TAT CAC TAA
```

C. Priano

FIGURE 16.4 Coding sequence of the human beta-globin gene.

Procedure 1a

For the first 120 nucleotides of the beta-globin gene, use the template at the end of this exercise to transcribe the DNA nucleotides into complementary mRNA nucleotides. Then, use the genetic code in Figure 16.2 to translate ribonucleotide codons into the first 40 amino acids of the beta-globin peptide.

1. On line 1, determine the complementary sequence of DNA nucleotides in the non-coding strand of the beta-globin gene.
2. On line 2, transcribe the mRNA from the **non-coding strand**. *Note, the non-coding strand is the template for transcription. What will happen if you use the coding strand as the template?*
3. On line 3, translate the first 40 amino acids of the encoded polypeptide.
4. Compare your sequence with your lab partners' sequences to check for accuracy.

Procedure 1b

1. Divide the remaining nucleotides of the beta-globin gene in Figure 16.4 among two to four lab partners.
2. Transcribe and translate the remainder of the beta-globin gene. Be sure to stay in the correct reading frame.

TABLE 16.1   Color code for amino acids.

ala	white	leu	light blue
arg	red	lys	brown
asn	black	met	tan
asp	solid yellow	phe	dark pink
cys	clear yellow	pro	light pink
gln	clear white	ser	dark green
glu	gray	thr	light green
gly	dark purple	trp	clear green
his	light purple	tyr	orange
ile	dark blue	val	clear blue

3. Using colored beads for amino acids, string together the 146 amino acid sequence (primary structure) onto the pipe cleaners. Use the color code in Table 16.1. Space the amino acids about 0.75 or 1.0 cm apart. Keep the amino acids equally spaced.

4. When all students have finished stringing together their amino acids, attach the pipe cleaners in order to construct the entire primary structure of beta-globin.

Answer the following questions:

1. What would happen if the sequence of the beta-globin gene were translated out of frame?

_____

_____

2. In your model of the primary structure of beta-globin, what does the pipe cleaner represent?

_____

_____

3. Is this sequence of amino acids a functional protein? Explain your answer.

_____

_____

ACTIVITY 2: FOLDING OF THE BETA-GLOBIN PEPTIDE

The amino acid sequence of the beta-globin polypeptide folds into eight right-handed alpha helices, designated A through H. These helical **domains** further fold into a tertiary conformation known as a globin fold. To visualize a right-handed helix, curl the fingers of your right hand with your thumb extended upward. As the helix spirals up, it curves in the direction of your fingers (see Figure 16.3, above). An alpha helix contains 3.6 amino acids per turn of the helix and is held stable by hydrogen bonds that form between the C=O of one amino acid and the N-H of the amino acid four places further along the chain.

Procedure

Simulate the eight alpha helices of the beta-globin peptide using the amino acid numbers in Table 16.2 as a guide. There are some non-helical segments in between successive helices. These are places where the polypeptide folds to form tertiary structure.

1. Using a drinking straw as a support, wind each helical domain into a right-handed helix around the length of the straw. Use four amino acids in each turn of the helix.

2. After you complete each helical domain, cut the straw to the size of that segment. Leave enough straw so that your helix does not fall off the end.

3. After all eight helices are complete, simulate the tertiary structure of the beta-globin peptide Refer to the schematic in Figure 16.5 to simulate the structure of the globin fold.

TABLE 16.2 Coordinates of beta-globin helices.

Helix	Amino Acid Positions	Amino Acids
A	4–18	thr-val
B	20–34	asn-val
C	35–41	tyr-phe
D	50–56	thr-gly
E	57–76	asn-ala
F	85–93	phe-cys
G	99–117	asp-his
H	123–143	thr-his

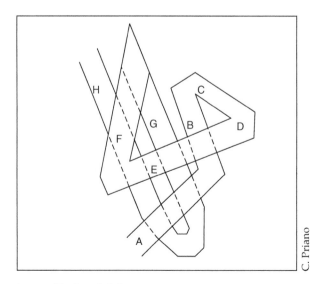

C. Priano

FIGURE 16.5 Tertiary structure of beta-globin.

III. COMPREHENSION CHECK

1. Describe the flow of genetic information in living organisms.

2. What is gene expression?

3. List three similarities and three differences between DNA synthesis and RNA synthesis.

4. Describe the following types of RNA and explain their functions in the cell:
   a. messenger RNA –

   b. ribosomal RNA –

   c. transfer RNA -

5. Describe why a polypeptide must fold to become an active protein.

6. In the model of the beta-globin gene, what does the straw simulate? Explain.

7. Describe the different types of secondary structures that can be adopted by proteins.

8. Explain how secondary structure is stabilized in a protein.

9. Describe the different kinds of interactions that help to stabilize the tertiary structure of a protein.

10. What is quaternary structure of a protein?

11. In the coding strand of the beta-globin gene, change nucleotide #20 from an A to a T. Discuss the potential consequences of this single nucleotide substitution.

Nucleotides 1-60

ATG GTG CAC CTG ACT CCT GAG GAG AAG TCT GCG GTT ACT GCC CTG TGG GGC AAG GTG AAC

Coding DNA strand →

non-coding strand →

1 ‖

Transcription →

mRNA →

2 ‖

Translation →

polypeptide →

3 ‖

Protein Folding →

mature protein

Nucleotides 61-120

GTG GAT GAA GTT GGT GGT GAG GCC CTG GGC AGG CTG CTG GTC TAC CCT TGG ACC CAG

Coding DNA strand →

non-coding strand →

1 ‖ ‖ ‖ ‖ ‖ ‖ ‖ ‖ ‖ ‖ ‖ ‖ ‖ ‖ ‖ ‖ ‖ ‖ ‖

Transcription →

mRNA →

2 ‖ ‖ ‖ ‖ ‖ ‖ ‖ ‖ ‖ ‖ ‖ ‖ ‖ ‖ ‖ ‖ ‖ ‖ ‖

Translation →

polypeptide →

3 ‖ ‖ ‖ ‖ ‖ ‖ ‖ ‖ ‖ ‖ ‖ ‖ ‖ ‖ ‖ ‖ ‖ ‖ ‖

Protein Folding →

mature protein

C. Priano

Use this Template to Express the First 40 Amino Acids of Beta-Globin

LABORATORY EXERCISE 17

RESTRICTION ENZYMES

After completing these laboratory exercises students should be able to:
1. Define restriction enzymes and DNA recognition sequences.
2. Explain how a restriction enzyme digests double-stranded DNA.
3. Construct a restriction map for a given DNA.
4. Recognize the use of restriction enzymes in biotechnology.

I. INTRODUCTION

Restriction enzymes identify specific **recognition sequences** in DNA and **digest** the DNA by cleaving phosphodiester bonds between successive nucleotides. These enzymes are **endonucleases** because they break bonds between internal nucleotides within the DNA molecule. Restriction enzymes are naturally produced by bacteria to destroy the DNA of invading **bacteriophages**. Bacterial DNA is not recognized or cleaved by restriction enzymes because of the presence of protective methyl ($- CH_3$) groups on nucleotides within recognition sequences. Hundreds of different restriction enzymes have been isolated and are important in many areas of biological research, including the sequencing of DNA genomes, the cloning of genes by recombinant DNA technology, and in analysis of DNA for forensic investigation. A restriction enzyme is named after the organism from which it is isolated, and is designated with three to four letters followed by a Roman numeral. For example, *Eco*RI is the name of a restriction enzyme isolated from the bacteria *Escherichia coli*.

All known restriction enzymes are categorized as either type I, type II, or type III (Table 17.1). A type I restriction enzyme cleaves double-stranded DNA in only one of the two complementary strands, about 1000–5000 base pairs away from the recognition sequence; a different enzyme must cleave the DNA in the second strand. A type II restriction enzyme identifies and cleaves both strands of a double-stranded DNA within the recognition sequence itself. A type III restriction enzyme recognizes two separate recognition sequences in a double-stranded DNA molecule, and the same enzyme cleaves both strands at a site far from the recognition sequences.

Type II restriction enzymes are commonly used in DNA technology. Their recognition sequences are **palindromic,** meaning that the nucleotide sequence reads the same from 5′ to 3′ in the two

181

TABLE 17.1 Types of restriction enzymes.

Type	Recognition site and cleavage site	Number of enzymes required for recognition and cleavage	Palindromic recognition sequence
I	Different	2	No
II	Same	1	Yes
III	Different	1	No

complementary DNA strands. For example, the restriction enzyme *Sma*I recognizes the DNA sequence 5'-CCCGGG-3', and cleaves the DNA between the C and G of each complementary DNA strand, as shown in Figure 17.1. Cutting straight through two complementary bases leaves each new double-stranded DNA fragment with a **blunt end**.

Other type II restriction endonucleases do not cut DNA straight across complementary nucleotides, but instead make a **staggered cut** that leaves some nucleotides unpaired on the ends of the new double-stranded DNA pieces. For example, the restriction enzyme *Eco*RI cleaves the DNA sequence 5'-GAATTC-3' between the G and the first A in each complementary strand, as shown in Figure 17.1. Notice that the resulting double-stranded DNA fragments have unpaired nucleotides on the 5' side of each complementary strand. These are called **5' overhangs**. Alternatively, the restriction enzyme *Sac*I cleaves its recognition sequence (5'-GAGCTC-3') between the T and the last C in each complementary strand. This cut leaves a **3' overhang** on the end of each resulting double-stranded DNA fragment.

The staggered ends of a restriction DNA fragment are also called **sticky ends**. Similar sticky ends can be generated in any other DNA molecule by cutting with the same restriction enzyme. The sticky ends can then be mixed and matched. In this way, DNA from different sources can be "cut and pasted" to create **recombinant DNA** molecules.

For any DNA, a diagram called a **restriction map** shows the location of specific restriction recognition sequences within the DNA molecule (Figure 17.2). A restriction map can be generated by digesting a DNA molecule with a given set of restriction endonucleases, then analyzing the sizes of the DNA fragments that are generated. DNA fragments of different sizes are separated using a technique called **gel electrophoresis** (Figure 17.3). In this process, negatively charged DNA is pulled toward a positively charged electrode through a soft gel medium. Smaller DNA fragments move

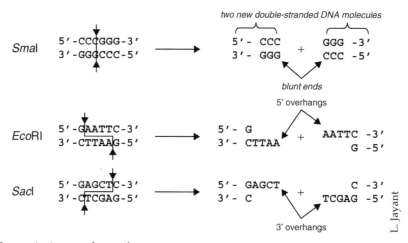

FIGURE 17.1 Type II restriction endonuclease

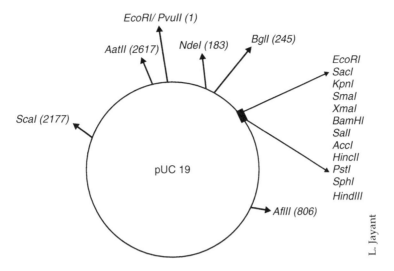

FIGURE 17.2   Restriction map of a circular DNA plasmid showing the locations of some major restriction sites.

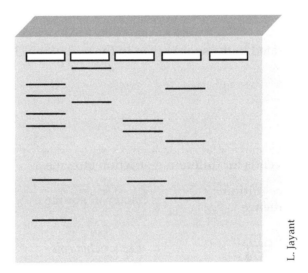

FIGURE 17.3   Cartoon of an agarose gel showing stained DNA fragments of various sizes after electrophoresis.

through a gel faster than larger fragments. The DNA fragments in the gel can be visualized by staining the DNA with a special dye. DNA fragment sizes can then be estimated and used to construct a restriction map.

II. LABORATORY ACTIVITIES

ACTIVITY 1: IDENTIFYING RESTRICTION SITES

Type II restriction enzymes have palindromic recognition sequences that are usually four, six, or eight nucleotides long. An enzyme that has a recognition sequence that is only four base pairs long is called a 4-cutter. Because there are only four different types of nucleotides in DNA, it can be estimated that a 4-cutter will cleave a DNA sequence $1/4^4$ times, or approximately once every 256 base pairs. An enzyme that has a recognition sequence that is eight base pairs long (an 8-cutter), would cut a DNA sequence approximately $1/4^8$ times, or once every 65,536 bases.

Approximately how many times will a restriction enzyme that is a 6-cutter cleave a DNA molecule that is 10,000 base pairs long?_____

Materials

colored pencils
scissors
DNA sequences 1 and 2 (found at the end of this Exercise)

Procedure

1. Examine DNA sequences 1 and 2 (for Activity 1).
2. Scan each DNA for the presence of palindromic sequences listed in Table 17.2 below.
3. Circle different recognition sequences in different colors. Use the same color to outline the enzyme in Table 17.2 that cuts this sequence. *(Note: Not all restriction recognition sequences will be present.)*
4. Predict whether there will be a 5' overhang, a 3' overhang, or a blunt end, and record in Table 17.2.
5. Use the scissors to simulate the restriction enzyme and cut the DNA exactly as shown in Table 17.2.
6. Stick the fragments generated in the space given in the end of the chapter.

Results

TABLE 17.2   Predicted DNA ends for different restriction enzymes.

Enzyme	Recognition sequence	Bacterial source	Ends
*Eco*RI	G AATTC CTTAA G	*Escherichia coli*	
*Pst*I	CTGCA G G ACGTC	*Providencia stuartii* 164	
*Hind*III	A AGCTT TTCGA A	*Haemophilus influenzae*	
*Bam*HI	G GATCC CCTAG G	*Bacillus amyloliquefaciens*	
*Sma*I	CCC GGG GGG CCC	*Serratia marcesans*	

ACTIVITY 2: GEL ELECTROPHORESIS SIMULATION

In gel electrophoresis, there is an inverse relationship between the log of the size of a DNA fragment and the distance traveled by the DNA through the gel matrix. A standard curve can be constructed by plotting the log of a fragment size (in base pairs) vs. distance migrated through the gel. The standard curve can then be used to estimate the sizes of unknown DNA fragments in a gel.

Materials

metric ruler
color pencils
scissors
glue
DNA sequences 3 and 4 at the end of this Exercise
Appendix 9

Procedure

1. Figure 17.4 shows a gel of a DNA molecular marker and the size of each DNA fragment.
2. Using a metric ruler, measure the distance travelled by each band from the top of the gel. Record the values in Table 17.3 below.

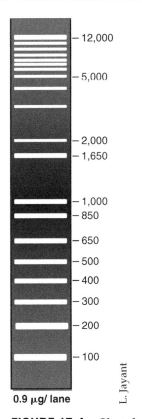

0.9 µg/ lane

L. Jayant

FIGURE 17.4   Sketch of molecular marker DNA after agarose gel electrophoresis.

TABLE 17.3   Migration distance of DNA marker fragments (cm).

Fragment length (bp)	Distance travelled (cm)
100	
200	
300	
400	
500	
650	
850	
1000	
1650	
2000	
3000	
4000	
5000	
6000	
7000	
8000	

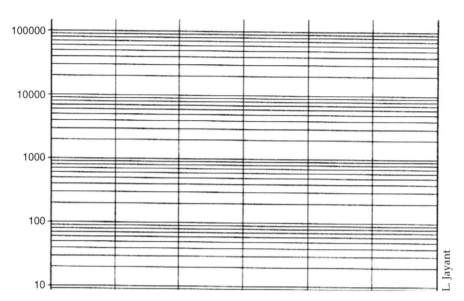

FIGURE 17.5 Standard curve.

3. Use the semi-logarithmic graph in Figure 17.5 to plot a standard curve. Put *Distance* in cm on the x-axis, and *Number of base pairs* on the y-axis. Plot the data for the DNA marker on the graph. Remember that the y-axis is a logarithmic scale. Make a small circle around each point and indicate the exact size of that fragment. Connect the points as best possible. Points between 500 and 3000 base pairs should make the straightest line.

4. Use sequences 3 and 4 (for Activity 2) at the end of this Exercise. Each member of a lab group will use their own DNA sequence to simulate a different restriction enzyme cut. Use scissors to cut each DNA with one of the following restriction enzymes: *Eco*RI, *Bam*HI, *Kpn*I, *Pst*I, and *Hind*III (refer to Appendix 9).

5. Record the fragment size for each digestion given below. Note: Each base pair in sequences 3 and 4 is equivalent to 50 base pairs of DNA in your digestion.
   • DNA fragments cut with *Eco*RI _____
   • DNA fragments cut with *Bam*HI_____
   • DNA fragments cut with *Kpn*I _____
   • DNA fragments cut with PstI_____
   • DNA fragments cut with *Hind*III_____

6. For each digestion, organize the resulting DNA fragments as they would appear as bands in separate lanes after gel electrophoresis. On a separate piece of paper, glue your bands in place. In Figure 17.6, draw each band on the gel according to size. Use the standard curve that you made to determine the distance traveled by each band (*e.g.,* If from the standard curve you determine that the distance travelled by a 4000 bp fragment is 5 cm, then place a 4000 bp fragment at a distance 5 cm from the well.)

ACTIVITY 3: RESTRICTION DIGESTION OF PLASMID DNA

For this activity, you can use the plasmid DNA that you isolated in Laboratory Exercise 15. Alternatively, plasmid pBR322 can be supplied at the discretion of the instructor. Note: Restriction enzymes have been supplied in colored microcentrifuge tubes. These are to be kept on ice at all times and are not to be discarded at the end of lab.

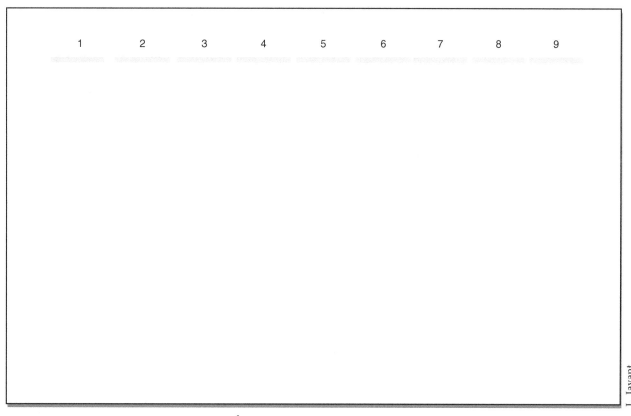

FIGURE 17.6   DNA bands in gel.

Materials

37°C water bath
microcentrifuge
microcentrifuge tube rack
microcentrifuge tubes
electrophoresis apparatus
power supply
pre-poured 1% agarose gels
micropipetters and tips
sterile water

plasmid DNA
restriction enzymes
TAE buffer
500 bp molecular marker (labeled MM)
NEB3 buffer (labeled NEB)
10x loading dye
GelRed stain
Appendix 7

Procedure

1. Label four microcentrifuge tubes as *"Eco*RI*", "Bam*HI*", "Eco*RI + *Bam*HI*"*, and "uncut plasmid". Keep these tubes on ice.

2. *If using your own isolated plasmid from Laboratory Exercise 15,* fill the first four tubes in the order shown in the Table 17.4 below.

3. *If using the supplied pBR322 plasmid,* make the following adjustment to Table 17.4 below: use only 2 μL of DNA per tube, and increase the amount of sterile water in each tube by 3 μL. Follow all other instructions in the table.

4. Incubate each digestion reaction in a 37°C water bath for 45 minutes.

5. Set up an agarose gel electrophoresis apparatus (see Appendix 7). Be sure that the gel is covered with TAE buffer and that all of the wells are filled.

6. After the incubation, push the contents of each tube to the bottom by spinning the tubes in a microcentrifuge for one minute. Return all tubes to a rack.

7. In a separate tube, add 5 µL of molecular marker plus 15 µL distilled water. This will be your DNA ladder.

8. To each of the five tubes (4 digestions and 1 molecular marker), add 3 µL of 10x loading dye.

9. Load 20 µL of the contents of tube 1 in the first well of the gel taking care not to spill the contents outside the well. (*Glycerol is present in the loading dye to increases the density of the sample and help to pull it down into the well*).

10. Load 20 µL of the remaining samples in the next four lanes. Note the order.

11. Attach the negative and positive electrodes making sure that the negative (black) electrode is closest to the wells.

12. Set the power supply unit at 100 V. Run the gel for 45 minutes or until the blue loading dye has migrated through two-thirds of the gel. Switch off the power and unplug the wires.

13. Carefully remove the gel and put it in GelRed stain for 15 minutes.

14. Pour the stain back into its container and rinse the gel in plain water for 5 minutes to destain.

15. Observe the destained gel on a UV light box. Be sure the shield on the UV light box is in place. DNA bands will fluoresce against a dark background.

16. Sketch the appearance of your gel in Figure 17.7 below.

TABLE 17.4 Setup for restriction digestion.

Tubes	Plasmid DNA (µL)	Sterile water (µL)	NEB buffer (µL)	EcoRI (µL)	BamHI (µL)	Total (µL)
1	5	12	2	1	0	20
2	5	12	2	0	1	20
3	5	11	2	1	1	20
4	5	15	0	0	0	20

Results

L. Jayant

FIGURE 17.7 Sketch of your gel.

ACTIVITY TO REINFORCE CONCEPTS

ACTIVITY 4: CONSTRUCT A RESTRICTION MAP

For this activity, refer to the following hypothetical experiment: Examine the circular plasmid pBMCC in Figure 17.8 below. Assume pBMCC is 11,400 base-pairs in size and that it was digested in separate reactions with either *EcoRI, Bam*HI, or *Hin*dIII. In addition, the plasmid was digested with different combinations of these restriction enzymes. For each reaction, the digested DNA fragments were subjected to agarose gel electrophoresis. The fragment sizes that were generated are shown in Table 17.5. Note that one kb is equal to 1000 base pairs. Create a restriction map for plasmid pBMCC.

Materials

Colored pencils

Procedure

1. Add the DNA fragment sizes generated from each digestion. Do they add up to 11.4 kb? _____

2. For each enzyme, place restriction sites on the plasmid in Figure 17.8. Hint: The enzyme *EcoRI* generates only one 11.4 kb fragment. This means that there is only one *EcoRI* recognition site in pBMCC. This site has already been placed on the plasmid diagram. Because the DNA is circular, this position is considered to be at 11.4 kb and at 0 kb.

3. Using the information in Table 17.5, determine the recognition sites for the remaining enzymes. Place each restriction site clockwise around the plasmid, starting at the *EcoRI* site. Keep in mind that some fragments were generated by more than one enzyme.

Results

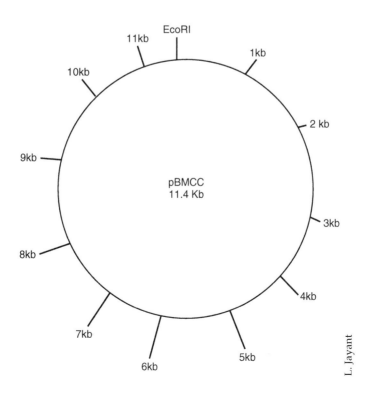

FIGURE 17.8   pBMCC.

TABLE 17.5   Restriction fragment lengths.

Reaction	Restriction enzyme(s) used	Sizes of resulting DNA fragments in kilobases (kb)
1.	EcoRI	11.4
2.	BamHI	4.4, 7.0
3.	Hind III	3.2, 8.2
4.	EcoRI + BamHI	7.0, 0.5, 3.9
5.	BamHI + HindIII	4.4, 2.0, 1.8, 3.2
6.	EcoRI + HindIII	3.2, 5.7, 2.5
7.	EcoRI + HindIII + BamHI	0.5, 2.0 , 3.2, 1.8, 3.9

Is another restriction map possible for pBMCC1? _____
If so, how does that map differ from the map you constructed?_____
_____

III. COMPREHENSION CHECK

1. What are restriction enzymes?

2. Why are restriction enzymes also called restriction endonucleases?

3. What kind of organism produces restriction enzymes? How do these organisms protect their own DNA from the restriction enzymes they produce?

4. A DNA is 10,000 nucleotides long. What is the frequency of nucleotide recognition sequences of 4, 6, and 8 nucleotides in this particular DNA? Explain your answer.

5. *Sac*I is a restriction endonuclease that recognizes the following sequence. What kind of overhang will this enzyme generate?

5'GAGCTC 3'
3' CTCGAG 5'

6. The linear map of a DNA fragment shown below has cleavage sites for the restriction enzymes *Bam*HI and *Eco*RI. In the accompanying diagram of an electrophoresis gel, indicate the positions at which bands would be found after electrophoresis if the DNA has been digested with:
   a. *Bam*HI alone   b. *Eco*RI alone   c. *Bam*HI and *Eco*RI together

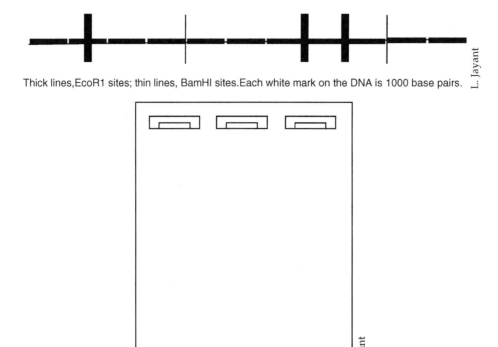

Thick lines,EcoR1 sites; thin lines, BamHI sites.Each white mark on the DNA is 1000 base pairs.

L. Jayant

7. Using the data from Activity 3, fill in the table below and construct a restriction map for the plasmid that you digested.

DNA fragment sizes		
	Enzyme	**Fragment size (kb)**
1.	*Eco*RI	
2.	*Bam*HI	
3.	*Eco*RI + *Bam*HI	
4.	Uncut Plasmid	

Restriction map

L. Jayant

FOR ACTIVITY 1

Sequence 1 restriction sites found: _____

In the space given below glue the fragments obtained after cutting the DNA with all the restriction enzymes found in the DNA sequence 1.

Sequence 2 restriction sites found: _____

In the space given below stick the fragments obtained after cutting the DNA with all the restriction enzymes found in DNA sequence 2.

SEQUENCE 1

5′ GAACATCGGATCCCACTCAGGCCCCGATGCTAGTACCATGGAATCCTGTGA 3′
3′ CTTGTAGCCTAGGGTGAGTCCGGGGCTACGATCATGGTACCTTAGGACACT 5′

SEQUENCE 2

5′ GAACATCCCGGGCCACTCAGGCCCCGATGCTACTGCAGTGGAATCCTGTGA 3′
3′ CTTGTAGGGCCCGGTGAGTCCGGGGCTACGATGACGTCACCTTAGGACACT 5′

FOR ACTIVITY 2

Assume every base pair (bp) in the sequence below is equivalent to 50 bp of DNA. Paste sequence 4 to the sequence 3 where it says 'paste'.

SEQUENCE 3

| 5′ GGCATTCCAC GGAATTCCAT CCCCTGGCGG GGCTGCAGTA CCGACTTATA 3′ | paste |
| 3′ CCGTAAGGTG CCTTAAGGTA GGGGACCGCC CCGACGTCAT   GGCTGAATAT 5′ |

SEQUENCE 4

5′ GCCAATATTG GGAATTCCAT CCCCTGGCGG GGCTGCAGTA CCGACTTATA 3′
3′ CGGTTATAAC CCTTAAGGTA GGGGACCGCC CCGACGTCAT   GGCTGAATAT 5′

DNA FINGERPRINTING

After completing these laboratory exercises students should be able to:

1. Comprehend the principle behind RFLP and DNA fingerprint technology in forensic analysis.
2. Apply the use of restriction enzymes for DNA fingerprinting.

I. INTRODUCTION

Even though all members of a species have essentially the same genetic makeup, slight differences in the DNA of their genomes, called **polymorphisms,** account for variations in phenotypes among individuals. Because restriction enzymes recognize specific DNA sequences, variations in DNA alter the restriction cutting patterns in different members of a species. Each individual therefore has a pattern of DNA restriction fragment lengths that are unique. This enables a method of genetic investigation known as Restriction Fragment Length Polymorphism, or **RFLP**, analysis.

RFLP is widely used in forensic DNA profiling. The human genome contains three billion nucleotides, 99.9% of which are identical among all people. The remaining 0.1% results in a distinct DNA restriction fragment pattern for every individual that is as unique as his or her fingerprint. For this reason, a person's RFLP pattern is commonly referred to as his or her **DNA fingerprint**. RFLP was the first DNA fingerprinting tool used for forensic DNA analysis. It has also been used to detect the presence of certain disease-causing genes.

RFLP technology involves cutting an individual's DNA with specific restriction enzymes and separating the different-sized DNA fragments by agarose gel electrophoresis. Because of the enormous amount of DNA in the human genome, restriction enzyme digestion of total genomic DNA yields far too many fragments to give a distinguishable pattern on a gel. In a technique known as **Southern blotting**, DNA restriction fragments that have been separated on an agarose gel are denatured to separate complementary strands, transferred to a nylon membrane, and then single-stranded DNA **probes** are used to locate desired genomic DNA sequences among the countless restriction fragments on the membrane. For DNA fingerprint analysis, the probes are relatively short pieces of DNA that are complementary to specific DNA sequences in the human genome where polymorphisms are common. In addition, each probe is attached to either a radioactive or a fluorescent **tag.** Because the DNA fragments on the nylon membrane have been denatured, the single-stranded DNA probes will specifically **hybridize** with their complementary counterparts in the genomic DNA fragments. When a probe hybridizes to a DNA fragment, its tag will emit a signal and "light up" the fragment on the membrane. Because restriction fragment lengths vary from one person to another, hybridization of the probes creates a unique pattern of different-sized bands for each individual, or a DNA fingerprint.

II. LABORATORY ACTIVITIES
ACTIVITY 1: DNA Fingerprint analysis
For the following procedure, refer to this hypothetical forensic investigation:

DNA was collected from a crime scene by forensic specialists. DNA samples were also obtained from two suspects that are being held in custody. At the crime lab, DNA from each source was digested with the restriction enzyme NotI. The resulting DNA fragments were subjected to agarose gel electrophoresis, then transferred onto a nylon membrane in a Southern blot procedure. A hypothetical cartoon of the nylon membrane with its denatured DNA fragments is shown in Figure 18.1 (only part of the DNA sequence for each fragment is shown). Several copies of two different single-stranded DNA probes can be found at the end of this Exercise.

Materials
colored pencils
scissors
glue
DNA probe sequences at the end of this Exercise

Procedure
1. Cut out each of the two probe sequences 1 and 2 at the end of this Laboratory Exercise. Color the back side of each probe fragment orange *(the orange color will represent the tag).*
2. Examine the cartoon of the nylon membrane in Figure 18.1. The arrangement of bands on the membrane is the same as on the agarose gel from which it was transferred, with the wells at the top. *Note that each DNA fragment shown is only a small part of a larger DNA molecule.*
3. Find a DNA sequence on the nylon membrane in Figure 18.1 that is complementary to either probe sequence 1 or 2. To simulate hybridization of the probe, glue the appropriate probe sequence to the DNA fragment so that the orange color is exposed. *Note: The complementary sequences do not have to match exactly. A probe will hybridize to a fragment even if there is as little as 50% complementarity with the DNA on the nylon membrane.*
4. Continue to search for sequences in the membrane (Figure 18.1) that are complementary to all or part of each probe.
5. In Figure 18.2, sketch the banding pattern that would emerge by RFLP analysis from the pattern of orange bands that you generated on your membrane.
6. Notice the different RFLP DNA fingerprint patterns for the different suspects and compare them to the pattern of the DNA found at the crime scene.

Results

```
┌──────────────────────────────┐   ┌──────────────────────────────┐   ┌──────────────────────────────┐
│                              │   │                              │   │                              │
└──────────────────────────────┘   └──────────────────────────────┘   └──────────────────────────────┘

GGCACCATCTACGGTACGATTGCA          CGTACGTACGTAGCATACATACATA         TTGCATTGCATTGCATTGCATTGCA

GGGGGCCCTATAGCGTGCATGCTT          CGTGACAACGTGACAGTGACAGTA          TTACTTCATTCATTCATCATTACATC

CGCGCCGCGCGATATAGTGTCACA          CCCCCGTGTACCCCAGTAGCAGTA          GACGACTTACGTAGCATGCTACGG

GACGACTTACGTAGCATGCTACGG          AAAACCGTCAGCTGACGCTAGCAG          GGCACCATCTACGGTACGATTGCA

GTCAGTCAGTCAGTCAGTCAGTCA          CGACGATGACGATCGCAGCTGACA          CGTCGTCGCTGCTGCTGCTGCTGCT

AAGTACGTACGTACGATCAGTACA          GGCACCATCTACGGTACGATTGCA          CACATCTGATCGATGACGTCGCTG

CCCGGTAGTGTTGGTTGGTTGTGT          CGCCGTACAGCTGACGCTACAGCT          CTTATTATTATTGATGCTGATCGTA

ACGACGGCACAGGCACAGGCAGC           GACGACTTACGTAGCATGCTACGG          CGCTGCTGACAGCTGACCGTGCAG

ACGACAGCCCAAAGTAACCCGTAA          GCTGCTAGCTGACGTGACTGACGT          CGGCTGATGGCGATGCCAGCTGAC

ACGTTGCAACGTTGCAACGTTGCA          CGTGATCGACCGTAGCAGCTGACT          GGCACCATCTACGCCGCATGGCAT

GACTTCAGTACTACGGTACGATTGC         CGTGACGACAGCTGACCGTAGCGA          GACGTAGCTGACAGCTGACCGTAG

GACGGTACGTATCAGGTGACTGGA          CGAGCTGAGCTGAGCAGCTGACCG          GACGACTTACGTAGCATGCTACGG

GACGGGTATTTGACGTAGCTAGCG          CGGTAGCGTAGCTCAGCTGACGTA          CGTTGCACGTAGCGATGATCGATG
```

FIGURE 18.1   Cartoon of DNA fragments in a nylon membrane for RFLP analysis.

L. Jayant

FIGURE 18.2   RFLP pattern for DNA fingerprint analysis.

Which samples match with the crime scene DNA? How would you interpret this information with respect to the crime? Explain your answer.

_____

_____

_____

_____

_____

_____

_____

_____

_____

_____

_____

_____

III. COMPREHENSION CHECK

1. What is the recognition sequence for the restriction enzyme *Not*I?

2. What kind of ends does the restriction enzyme *Not*I create in the digested DNA?

3. *Not*I is an 8-cutter. How many times will it cut a DNA that is 3 billion base pairs long? Explain.

PROBE SEQUENCES TO CUT OUT FOR ACTIVITY 1

Probe 1

CCGTGGTAGATGCCATGCTAACGT

CCGTGGTAGATGCCATGCTAACGT

CCGTGGTAGATGCCATGCTAACGT

CCGTGGTAGATGCCATGCTAACGT

CCGTGGTAGATGCCATGCTAACGT

CCGTGGTAGATGCCATGCTAACGT

CCGTGGTAGATGCCATGCTAACGT

CCGTGGTAGATGCCATGCTAACGT

CCGTGGTAGATGCCATGCTAACGT

Probe 2

CTGCTGAATGCATCGTACGATGCCC

CTGCTGAATGCATCGTACGATGCCC

CTGCTGAATGCATCGTACGATGCCC

CTGCTGAATGCATCGTACGATGCCC

CTGCTGAATGCATCGTACGATGCCC

CTGCTGAATGCATCGTACGATGCCC

CTGCTGAATGCATCGTACGATGCCC

CTGCTGAATGCATCGTACGATGCCC

CTGCTGAATGCATCGTACGATGCCC

L. Jayant

LABORATORY EXERCISE 19

DNA CLONING AND BACTERIAL TRANSFORMATION

After completing this exercise, students should be able to do the following:
1. List the requirements of a good plasmid expression vector.
2. Use the appropriate restriction enzymes to clone a DNA fragment into a plasmid.
3. Transform plasmid DNA into bacteria.
4. Explain the use of an antibiotic and an inducing agent in a transformation procedure.

I. INTRODUCTION

Gene cloning is a process in which a gene of interest can be copied, or cloned. A foreign DNA containing a gene of interest can be inserted into a bacterial DNA called a **plasmid** by "cutting and pasting" the different DNA pieces (Figure 19.1). A plasmid is a small circular DNA molecule that exists separate from the bacterial chromosome. In gene cloning, a foreign gene is first "cut" from its own genome using special enzymes called **restriction endonucleases**. A bacterial plasmid is also cut in the same way to linearize the plasmid DNA. The foreign DNA fragment and the linear plasmid DNA are mixed together in the presence of another enzyme, **DNA ligase**, which "pastes" the foreign gene into the plasmid. When cutting and pasting DNA, both the foreign DNA and the plasmid must be digested with the same restriction endonuclease so that they will have compatible or matching ends.

Once the foreign DNA is pasted or **ligated** into the plasmid, the plasmid is a **recombinant DNA vector** for the gene, which means that it can carry the foreign gene into a bacterial **host**. A good plasmid vector must have an **origin of replication** *(ori)* that ensures proper replication of the plasmid within the bacterial host. It must also carry a gene for **antibiotic** resistance. An antibiotic is a substance that destroys or inhibits the growth of bacteria. A bacterium that carries a DNA plasmid acquires resistance to a particular antibiotic. When the host bacterium reproduces, bacterial DNA polymerase replicates the plasmid DNA along with the host cell DNA. Each daughter cell then receives a copy of the plasmid. This is the basis for the rise of antibiotic resistant strains of bacteria, such as methicillin-resistant *Staphylococcus aureus*, or MRSA.

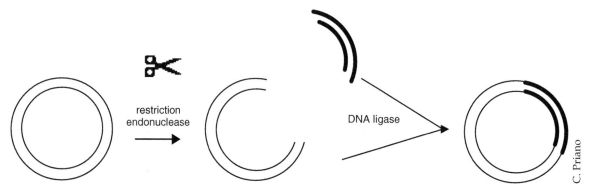

FIGURE 19.1 Restriction endonuclease digestion and ligation. "Cutting and pasting" of a foreign gene into a plasmid.

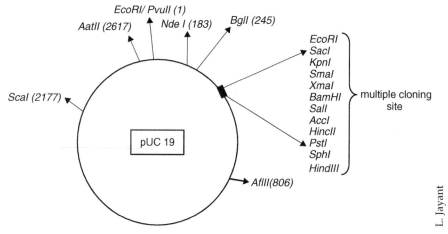

FIGURE 19.2 Map of a plasmid showing restriction endonuclease sites and a multiple cloning site.

Bacterial **transformation** is the process by which bacterial cells take up foreign DNA molecules. In the presence of an antibiotic, bacteria can be induced to take up plasmid DNA that carries a gene for resistance to the antibiotic. Bacteria must first be made **competent**, or able, to take up foreign DNA through the plasma membrane. This is done by exposing dividing bacteria to ice-cold calcium chloride ($CaCl_2$), then raising the temperature of the bacteria to 42°C, a process known as **heat shock**. Heat shock turns on a set of bacterial "heat shock" genes that allow the bacteria to tolerate the higher temperature. Plasmid vectors can be maintained in bacteria by growing the bacteria in the presence of the appropriate antibiotic. This ensures that only bacteria that contain the plasmid will survive in the presence of the antibiotic. In this way, the gene of interest carried on the plasmid vector is copied, or cloned, within the growing bacteria.

To ensure that the gene of interest is transcribed by bacterial RNA polymerase, the cloned gene must be under the control of a transcriptional **promoter**. Commercially sold plasmid **expression vectors** have a specific prokaryotic promoter already in place so that a foreign gene can be easily inserted and expressed. Such promoters are directed into a **multiple cloning site** that houses several restriction recognition sites (Figure 19.2). Having many restriction sites conveniently provides a wide range of options for choosing the appropriate restriction endonuclease needed to clone a gene. In some vectors, a promoter is under the control of a control element called an **operator** that regulates transcription. In these plasmids, transcription requires the presence of an **inducing agent**. For example, the *lac* operator prevents transcription under normal conditions. However, in the presence of lactose, transcription of messenger RNA (mRNA) for a cloned gene can be induced. Because the genetic code is universal, the translational components of the bacteria can express the foreign mRNA and generate encoded proteins.

Gene cloning has widespread use in research for examining genomic DNA, gene products, and the mechanisms that regulate gene expression in prokaryotic and eukaryotic organisms. For decades, such technology has provided important applications in medicine, for example in the production of human insulin to treat type 1 diabetes mellitus.

II. LABORATORY ACTIVITIES
ACTIVITY 1: CONSTRUCTION OF THE pGLO PLASMID USING A PAPER MODEL

In a living organism, all genes are not expressed at the same time. Selective regulation of gene expression is important for cellular specialization, for development, and to prevent unnecessary energy use in overproduction of proteins. In protein synthesis, formation of each peptide bond requires four ATPs. Therefore, gene expression is tightly regulated. Depending on the cell type, only genes for necessary products are expressed.

Most organisms regulate gene expression at the level of transcription. In prokaryotes, some genes that are involved in a single metabolic pathway are clustered together in groups called **operons.** All genes of an operon are regulated by the same promoter-operator region. For example, bacterial genes involved in lactose metabolism are located in the *lac* operon, and those involved in arabinose metabolism are grouped in the arabinose operon.

Figure 19.3 illustrates the arabinose operon. Clustered in this operon are three genes, *araB, araA,* and *araD.* These genes encode enzymes required for the breakdown of the sugar arabinose. To transcribe these genes, RNA polymerase binds to a promoter region *(araBAD)* on the 5′ side of the operon. A DNA **repressor** protein, *araC,* is involved in regulation of this operon. In the absence of arabinose, *araC* binds to the promoter and prevents the attachment of RNA polymerase. If arabinose is present, it will associate with *araC* and remove it from the promoter. RNA polymerase can then bind to the promoter and initiate transcription of all genes in the operon. Once arabinose is completely broken down by the enzymes produced by the operon, *araC* will bind again to the promoter and prevent further transcription of the arabinose operon.

The plasmid pGLO uses elements of the arabinose operon to regulate expression of an inserted gene. The inserted gene you will use in this exercise expresses a protein called **GFP** or **Green Fluorescent Protein**. This protein is made by jellyfish and will glow in the presence of ultraviolet (UV) light. In pGLO, expression of the GFP gene is under the control of the *araBAD* promoter. Transcription of the GFP gene is induced when arabinose (the inducing agent) is present. The plasmid also codes for the *araC* repressor protein and carries a gene for ampicillin resistance (Amp^r) (Figure 19.4).

A good expression vector should have the following features:

- An origin of DNA replication
- An antibiotic resistance gene
- A bacterial promoter that will be recognized by a bacterial RNA polymerase
- A regulating mechanism to regulate expression of the cloned genes
- A multiple cloning site next to the promoter to be able to insert the gene of interest

Answer the following:

Why does a plasmid vector need a gene for antibiotic resistance? What would happen if it didn't have such a gene? Explain your answer in complete sentences. _____

_____

_____

_____

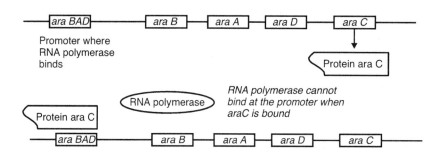

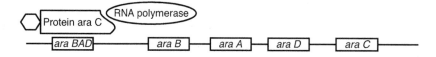

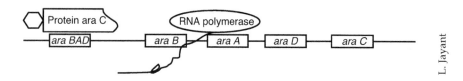

FIGURE 19.3 The arabinose operon.

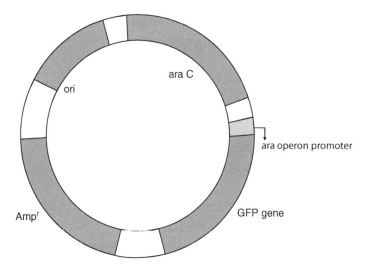

FIGURE 19.4 The plasmid pGLO.

Materials

colored pencils   glue
scissors   Appendix 9

Procedure

1. To simulate the plasmid pGLO, cut out each strip of the pGLO plasmid DNA found on the last page of this Exercise. Glue the right side of strip 1 to the left side of strip 2. Continue to glue each line together in succession. To circularize your plasmid, glue the end of the DNA at the right side of strip 7 to the beginning of strip 1.

2. Observe your plasmid. Color the *ori* or origin of replication blue.

3. Find the antibiotic resistance gene and color it orange. Find the promoter and color it red. Find the *araC* gene and color it purple.

4. Cut out the chromosomal DNA containing the GFP gene on the last page of this exercise.

5. Refer to Appendix 9 to find restriction endonucleases that will cut the DNA on either side of the GFP gene.

6. Find the same restriction recognition sites in the Multiple Cloning Site of the pGLO plasmid. (*Hint: The Multiple Cloning Site is on strip 5 of your plasmid.*)

7. Using scissors to simulate restriction enzymes, cut the GFP gene from the chromosomal DNA and cut the pGLO plasmid in the Multiple Cloning Site, using the appropriate compatible restriction enzymes.

8. Using glue to simulate DNA ligase, ligate the GFP gene into the plasmid to contruct the recombinant plasmid pGLO/GFP.

9. Color the inserted GFP gene green. Paste your plasmid in Figure 19.5.

10. Check to see if your pGLO has all the elements of a good expression vector.

Answer the following questions:

1. What restriction endonuclease(s) did you use?_____

2. To what antibiotic is your plasmid resistant?_____

3. What is the name of the promoter?_____

4. Where did you cut the plasmid to insert the GFP gene?_____

5. What enzyme did you use to "paste" the DNA together?_____

Results

L. Jayant

FIGURE 19.5   Your plasmid (glue in place).

ACTIVITY 2: TRANSFORMATION OF pGLO INTO A BACTERIAL HOST.

In this activity, we will transform bacteria with the DNA plasmid pGLO. Transformation is the result of a foreign DNA being expressed in a host cell. In a simple procedure, a bacterial plasma cell membrane is made **permeable** to a piece foreign DNA that contains a gene (Figure 19.6). In this procedure, ice-cold calcium chloride ($CaCl_2$) is used to neutralize negative charges on phospholipid heads of the plasma membrane, so that the membrane does not repel the negatively charged DNA. A rapid increase in temperature to 42°C, or heat shock, helps to increase the uptake of DNA by a cell. It is

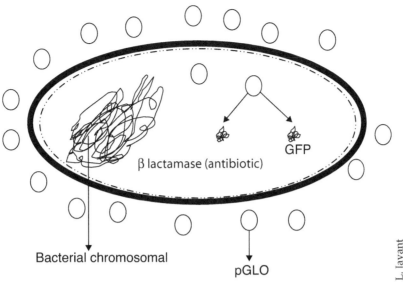

FIGURE 19.6   Bacterial chromosomal DNA

believed that the heat shock disorients the membrane lipids, creating tiny pores for DNA to enter. The duration of heat shock is crucial. While a longer heat shock could kill the bacteria, a shorter heat shock might not allow the cells to take up the DNA plasmid. Following heat shock, the cells are given a 15-minute incubation period during which time they begin to synthesize antibiotic resistant enzymes that will allow the cells to grow in the presence of antibiotic.

You will transform the bacteria *E. coli* with a recombinant pGLO plasmid that has the GFP gene next to the arabinose operon promoter. This is the recombinant plasmid that you simulated in Activity 1. Transformed host bacteria will be grown on agar media that contains the antibiotic ampicillin and arabinose as the gene-inducing agent.

Materials

37°C incubator	competent *E. coli* HB101
UV penlight	pGLO plasmid
water bath set at 42°C	50 mM CaCl2 solution (pH 6.1)
micropipeterer (P20)	sterile LB agar plates (LB, LB/amp, and LB/amp/ara)
pipette tips (20 µl)	sterile LB medium
2 mL microcentrifuge tubes	sterile loops
	sterile transfer pipettes

Procedure

Note: Maintain sterile conditions. Leaving the tubes open or touching them with non-sterile objects will lead to contamination and affect transformation results.

1. Label one microcentrifuge tube +pGLO and label a second tube –pGLO.

2. To each tube, add 250 µL of sterile ice cold CaCl$_2$ using a sterile transfer pipette. Keep the tubes on ice.

3. To each tube, add one single colony of bacteria using a sterile loop provided. Immerse the loop into the CaCl$_2$ solution. Gently and thoroughly mix the cells making sure there are no clumps. Return the tubes to the ice. **Discard the loops in a biohazard safety bag.**

4. Look at the tube of the plasmid pGLO under UV light. Note your observations: _____

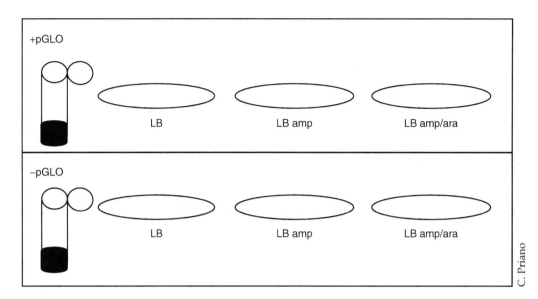

FIGURE 19.7 Plating bacterial transformants.

5. Using a micropipetter, add 10 µL of pGLO into the tube labeled +pGLO. Add 10 µL of sterile water into the tube labeled –pGLO. *This tube is the* _____. Return the tubes to ice.

6. Incubate the tubes on ice for 10 minutes.

7. After 10 minutes, carry your ice bucket to the 42°C water bath and transfer both tubes directly from the ice into a water bath. Incubate each tube at 42°C for exactly 50 seconds, then return immediately to the ice.

8. Leave tubes in ice for an additional 2 minutes.

9. Add 250 µL of sterile LB medium to both the +pGLO and –pGLO tubes and incubate at room temperature for 15 minutes.

10. While the tubes are being incubated at room temperature, label one of each type of agar plate +pGLO, and one of each type as –pGLO, as shown in Figure 19.7. Label the agar plates on the bottom; do not label the lids. (Agar plates are LB, LB/amp, and LB/amp/ara.)

11. After the 15-minute incubation, divide the contents of the +pGLO tube onto the three +pGLO agar plates as shown in Figure 19.7 (approximately 150 µL each onto LB, LB/amp, and LB/amp/ara).

12. Repeat step 11 with the contents of the –pGLO tube and the –pGLO agar plates.

13. Incubate the agar plates in a 37°C incubator for 24–37 hours.

14. After incubation, record your results in Table 19.1. Illuminate the plates with the UV penlight to observe colonies that are expressing the green fluorescent protein.

Results

TABLE 19.1 Number of bacterial transformants obtained.

Plates	LB (total)	LB amp (total)	LB amp + ara (total)	LB amp + ara (green under UV)
+pGLO				
–pGLO				

III. COMPREHENSION CHECK

1. What are the steps for cloning a foreign gene into a plasmid?

2. What are the requirements for a good expression plasmid vector?

3. The bacterial *lac* operon contains genes that code for enzymes needed to metabolize lactose. The operon has a promoter and operator. You want to clone these genes into a plasmid vector. You cut the part of the operon containing only the genes, insert them into a plasmid, then transform bacteria with this recombinant plasmid. Will the transformed bacteria express these proteins? Why or why not? Explain your answer.

4. Why do we use $CaCl_2$ for bacterial transformation?

5. What is the purpose of heat shock during transformation?

6. You cut a gene from a chromosome using the restriction enzymes *Eco*RI and *Bgl*II. You cut a plasmid with *Pst*I and *Hind*III. Is it possible to insert the gene segment into this plasmid? Why or why not?

7. Most plasmid vectors code for an antibiotic resistant gene. Explain the need for this gene.

8. Bacterial clones that carry a plasmid are maintained in a medium that contains an antibiotic to which the bacteria are resistant. What will happen if the antibiotic is removed from the medium? Explain your answer.

pGLO plasmid DNA

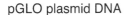

1

```
5' GGACTAAACCCTAXXXXXXORIXXXXXXXXXXGAGGCATTTA 3'
3' CCTGATTTGGGATXXXXXXORIXXXXXXXXXXCTCCGTAAAT 5'
```

2

```
5' GACCCTACTGGCAGTCAGTCTACGGAATTCGCTAGGCTACCT 3'
3' CTGGGATGACCGTCAGTCAGATGCCTTAAGCGATCCGATGGA5'
```

3

```
5' XXXXXXXXXXXARACXXXXXXXXXXXXXXXXXXXXXXXXXXX 3'
3' XXXXXXXXXXXARACXXXXXXXXXXXXXXXXXXXXXXXXXXX 5'
```

4

```
5'GATCCAGTXXXXPROMOTERXXXXXXXXXXXXXXXXXXXXX 3'
3' CTAGGTCAXXXXPROMOTERXXXXXXXXXXXXXXXXXXXXX 5'
```

5

```
5' GGCCGGATCCGAATTCCAAGCTTGGTACCAAGCTTCCCGGG 3'
3' CCGGCCTAGGCTTAAGGTTCGAACCATGGTTCGAAGGGCCC 5'
```

6

```
5' GCTAGCCTGCAGGTCGACCTCGAGGACTCCGACCGTCGCTA 3'
3' CGATCGGACGTCCAGCTGGAGCTCCTGAGGCTGGCAGCGAT 5'
```

7

```
5' CCGTAATTCCCCGGGXXXXAMPICILLINXXXXXXXXXXXXX 3'
3' GGCATTAAGGGGCCCXXXXAMPICILLINXXXXXXXXXXXXX 5'
```

Chromosomal DNA with GFP gene

```
5' CAGCGGTACCATCCGGXXXGFPXXXXXXXXXXXXCTGCAGCCC 3'
3' GTCGCCATGGTAGGCCXXXGFPXXXXXXXXXXXXGACGTCGGG 5'
```

L. Jayant

LABORATORY EXERCISE 20

GENETICS

After completing this exercise, students should be able to do the following:
1. Explain how genetic traits are expressed and inherited.
2. Comprehend the relationship between alleles and homologous chromosomes.
3. Identify chromosomal abnormalities by examining a karyotype.
4. Distinguish between phenotype and genotype for common dominant and recessive traits.

I. INTRODUCTION

A. Classical Genetics

Genetics is the branch of biology concerned with the transmission of inherited traits. A **gene** can be defined as a unit of inheritance, as a **locus** on a chromosome that corresponds to a characteristic, or as a segment of DNA that carries information, usually for making a protein. Diploid cells of sexually reproducing organisms contain two complete sets of DNA, and therefore two copies of every gene. Variable forms of a single gene are known as **alleles**. Whereas the gene holds information for the expression of a characteristic, an allele is responsible for a specific trait, or a distinguishing feature of a characteristic. For example, in sweet peas, a gene for flower color, the characteristic, might have alternative forms or alleles that result in different color traits, perhaps purple or white. An individual that has two copies of the same allele for a gene *(two purple alleles)* is **homozygous**, whereas an individual that has two different alleles for a gene *(one purple and one white allele)* is **heterozygous**.

Alleles are often either **dominant** or **recessive**. An allele is dominant if it is expressed, or seen, in an individual that is either homozygous or heterozygous. An allele is recessive if it is only expressed when the individual is homozygous for that allele. Dominant alleles are often designated with a capital letter, whereas recessive alleles are given the same letter, but in lower case *(purple flower color, P is dominant to white flower color, p)*. The actual genetic makeup of an individual is its **genotype** and is designated by combining the abbreviations for the alleles *(PP is a homozygous dominant genotype, pp is homozygous recessive genotype, and Pp is a heterozygous genotype)*. The actual appearance of the traits in the individual is its **phenotype** *(flower color is either purple or white)*.

B. Chromosomes and Inheritance

The basis for genetic inheritance lies in the behavior of chromosomes during cell division. In meiosis, separation of homologous pairs of chromosomes causes the allele pairs for each gene to be segregated as well. Each haploid gamete receives only one allele from the parent cell. Upon fertilization, alleles are again paired, but in different combinations. The mixing and matching of alleles leads to most of the variation present in sexually reproducing populations. Because of the way alleles segregate and reunite, patterns of inheritance can be traced through generations and can also be predicted for future generations. For example, we can use methods such as **pedigree analysis** to track the history of how a genetic trait arose in an individual, or we can use **Punnett square** analysis to predict the probability of a genetic trait arising in a future generation.

C. Human Genetics

Human genetics deals in part with the inheritance patterns of human genetic traits. A human diploid cell has two complete sets of chromosomes: 22 pairs of **autosomes** and two **sex chromosomes**, which are either two X chromosomes in females, or one X and one Y chromosome in males. Figure 20.1 is an example of a human **karyotype**, a display of one's chromosomes according to size and centromere position. More than 21,000 genes that make up the human **genome** are distributed among each set of these chromosomes. The inheritance patterns of human genetic traits differ depending on whether a gene locus is on an autosome (autosomal inheritance) or on a sex chromosome (sex-linked inheritance).

In the case of genetic disease, methods are often employed to analyze transmission of genetic disorders, such as Tay Sachs disease or cystic fibrosis. In many cases, a person's DNA can be analyzed directly to determine the presence or absence of a defective allele for such a gene. Risk factors can then be taken into consideration when planning a family. In addition, in high-risk situations, DNA testing can be performed **prenatally** to determine if a fetus bears a particular genetic disorder.

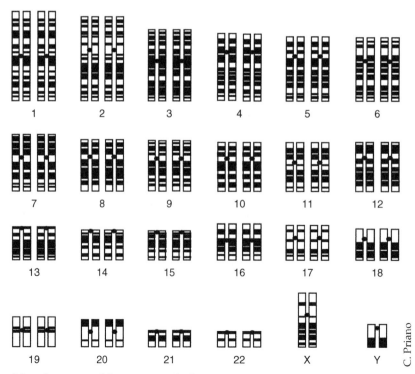

C. Priano

FIGURE 20.1 Graphic of a normal human male karyotype.

TABLE 20.1   Common human aneuploidies

Aneuploidy	Condition	Description
Trisomy 13	Patau Syndrome	severe nervous system, muscular, and organ defects; death usually occurs in early infancy
Trisomy 18	Edwards Syndrome	mental and physical defects; death often in infancy
Trisomy 21	Down Syndrome	mild to severe physical and mental defects; survival rates high
Monosomy X	Turner Syndrome	all female; short stature; undeveloped secondary sex characteristics; infertility
XXY	Klinefelter Syndrome	all male; few detectable symptoms; reduced fertility

An abnormality in chromosome number or structure can also contribute to genetic defects. An abnormal number of chromosomes is known as **aneuploidy**. Aneuploidies result from a defect in cell division known as **nondisjunction**, in which chromosomes fail to separate properly. Nondisjunction during either meiosis I or meiosis II can result in gametes that have either too many or too few chromosomes. For example, one extra copy of chromosome 21, or **trisomy** 21, results in Down syndrome, a genetic condition in which an individual has a variety of physical and medical disorders that can range from mild to severe. For most human aneuploidies, the chromosomal imbalance results in spontaneous abortion of a fetus early in pregnancy. Aneuploidies can involve either autosomes or sex chromosomes (Table 20.1). Another type of chromosomal abnormality is known as a **translocation**, in which there is a rearrangement of parts of different, nonhomologous chromosomes. Translocations can lead to deletion of parts of genes, or fusion between different genes. Aneuploidies and translocations can be observed by **cytogenetic** analysis, in which a person's chromosomes are directly observed and analyzed.

II. LABORATORY ACTIVITIES
ACTIVITY 1. HUMAN KARYOTYPE ANALYSIS

A **karyotype** is a photograph of an individual's chromosomes displayed in an array according to size and centromere position. Karyotypes are constructed for the purpose of chromosomal analysis. A karyotype can be prepared for an individual using the white blood cells taken from a simple blood sample. For prenatal diagnosis, a special procedure called **amniocentesis** is employed to determine whether any chromosomal abnormality is present in the cells of a fetus. This procedure is often performed on pregnant women over the age of 35, when there is an increased risk of **aneuploidy** (abnormal number of chromosomes) in the fetus. During amniocentesis, a small amount of **amniotic fluid** that surrounds the fetus is removed with the aid of **ultrasonography**. A needle is inserted into the mother's abdomen and a small amount of amniotic fluid is safely removed from the uterus. This fluid contains fetal cells, mostly from the skin of the developing fetus. These cells are then used to analyze the chromosomes of the fetus.

Once the cells are collected, they are stimulated to undergo cell division using a chemical substance called a **mitogen**. After a period of cell growth, cells are arrested in metaphase of mitosis by adding another chemical substance called **colchicine**, which freezes the mitotic spindle. These cells are then concentrated and put into a hypotonic solution, which causes the cells to swell. When dropped onto a glass slide, the swollen cells break open and the chromosomes "spill" out onto the slide. This creates a **chromosome spread** for each metaphase cell, which can be viewed with a light microscope. The chromosomes are stained with a dye *(for example, Giemsa stain)* that dyes the chomosomes in a characteristic **banding pattern**. These banding patterns *(for example, G-bands from Giemsa stain)* help to identify each of the different homologous chromosome pairs. Stained chromosomes from a spread are photographed under microscopy. Photographs are enlarged, and chromosome pairs are matched, cut out, and arranged according to size and centromere position into a karyotype.

In this activity, you will prepare a karyotype of G-banded chromosomes from a chromosome spread that has been stained with Giemsa. After your karyotype is complete, you will identify the sex of the individual, whether the karyotype is normal or, if not, what chromosomal aneuploidy is present. Work in groups of two students per chromosome spread. Each group of students should have a different karyotype.

Materials

photograph of a chromosome spread
pencil
scissors

glue stick
graphic of a human karyotype
human karyotype form

Procedure

1. Obtain a chromosome spread and the karyotype form found at the end of this Exercise.

2. Count the chromosomes on your chromosome spread.

3. Circle each homologous pair that you can identify by matching the size, centromere position, and banding pattern. Keep track of each pair that you identify.

4. Cut out each chromosome pair.

5. Using the graphic of the finished karyotype (Figure 20.1) as a guide, place each of your homologous pairs in the appropriate position on the human karyotype form provided. When you are sure of the position for each pair of chromosomes, glue in place.

6. Determine the following information and complete the karyotype form:

   a. What is the total number of chromosomes?

   b. What is the sex of the individual?

   c. Is this a normal karyotype?

   d. If not, what is the chromosomal abnormality?

ACTIVITY 2. DOMINANT AND RECESSIVE GENETIC INHERITANCE

Below is a list of human phenotypes that follow simple dominant/recessive inheritance patterns. Traits are distinguishing features of phenotypic characteristics. Dominant traits are always expressed, whether an individual is homozygous for the trait or heterozygous. Recessive traits are only expressed if the individual is homozygous for the trait.

Procedure 2a

Fill in your phenotypes for the characteristics listed in Table 20.2. Alleles for each trait are represented by either capital letters for dominant traits or lowercase letters for recessive traits. If your phenotype for a characteristic is recessive, then you are homozygous recessive and you must have two recessive alleles in your genotype. If your phenotype is dominant, then you can be either homozygous dominant or heterozygous. In this case, you will list two possible genotypes.

Procedure 2b

Determine the frequency of phenotypes in your class by scoring the class data for the characteristics listed in Table 20.3. The frequency of a phenotype is determined by dividing the number of students that have that phenotype by the total number of students in the class.

TABLE 20.2   Simple dominant/recessive inheritance

Characteristic	Dominant trait	Recessive trait	Genotype(s)
Interlocked fingers	thumbs cross left over right (*L*)	thumbs cross right over left (*l*)	
Earlobe shape	unattached (*E*)	attached (e)	
Hair on back of hand	hair present (*H*)	not present (*h*)	
Eye shape	almond (*A*)	round (*a*)	
Dimples	dimples present (*D*)	not present (*d*)	
Eyebrows	separated (S)	not separated (*s*)	
Hairline	widow's peak (*W*)	straight (*w*)	
Thumb shape	straight (S)	curved (*s*)	
Cleft chin	cleft absent (C)	cleft present (*c*)	

TABLE 20.3   Frequencies of phenotypes in your class

Characteristic	Number of students with recessive phenotype	Frequency of recessive phenotype	Number of students with dominant phenotype	Frequency of dominant phenotype
Interlocked fingers				
Earlobe shape				
Hair on back of hand				
Eye shape				
Dimples				
Eyebrows				
Hairline				
Thumb shape				
Cleft chin				

Answer the following questions:

1. How does the frequency of the recessive phenotype for a characteristic compare with the frequency of the genotype for that characteristic?

_____
_____
_____
_____
_____
_____
_____
_____
_____
_____
_____
_____
_____
_____
_____
_____
_____
_____
_____
_____
_____
_____

2. Is it possible to determine the genotype of students that express the dominant phenotypes for any of the characteristics listed above? Explain.

_____
_____
_____
_____
_____
_____
_____
_____
_____
_____
_____
_____
_____
_____
_____
_____
_____
_____
_____
_____
_____
_____
_____
_____

III. COMPREHENSION CHECK

1. Explain the difference between a gene and an allele.

2. Explain the terms homozygote and heterozygote.

3. How can a defect in a chromosome affect the genetic balance in an individual?

4. What kinds of genetic information can be determined by karyotype analysis?

5. What is the usefulness of karyotype analysis?

6. If 40% of students in a class have dimples, and 50% of these students have the genotype DD, what are the phenotypic and genotypic frequencies for dimples in the class?

KARYOTYPE ANALYSIS FORM

1	2	3	4	5
	A			B
6	7	8	9	10	11	12

6	7	8	9	10
			C	

11	12

13	14	15	16	17	18
	D			E	

19	20	21	22	X	Y
F		G		sex chromosomes	

Number of chromosomes _____   Sex _____   Disorder _____

C. Priano

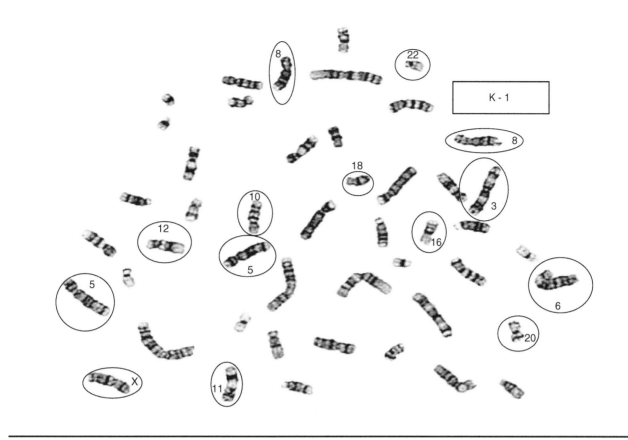

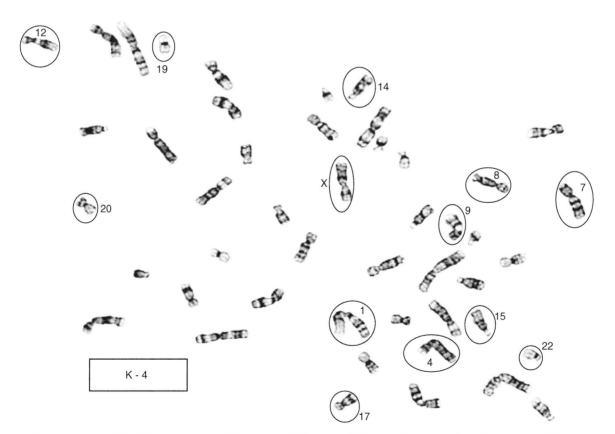

Univ. Wisconsin

LABORATORY RULES AND SAFETY INSTRUCTIONS

Safety rules are of the utmost importance when working in a laboratory. These instructions are meant for the protection of everyone working in the lab, as well as for those who will be using the lab after you:

1. Food, chewing gum, and beverages (including water) are prohibited in the lab.
2. Aisles are to be kept clear at all times. Book bags, coats, etc. are to be kept in a cabinet or on a hook. Empty chairs are to be pushed under the tables.
3. Do not taste or inhale any substances in the lab. Avoid contact with skin and eyes and inform the instructor of any spills.
4. Do not pipette any substance by mouth. Use an aspirator bulb or automatic pipetter.
5. Do not directly touch any hot materials. Use tongs or an oven mitt.
6. Keep loose hair tied back at all times.
7. If you are using biohazardous materials, be sure to dispose of them in the appropriate containers.
8. Do not dispose of garbage in any of the sinks; use the trash cans.
9. Broken glass is to be discarded in a glass disposal bin, not with the regular trash.
10. Before using any equipment in the lab, wait for your instructor to demonstrate the proper handling and use. Never touch any equipment that you do not know how to operate.
11. You are responsible for any lab equipment that you break or damage.
12. Each student is responsible for the table at which he or she is working. Before leaving the lab, all supplies are to be properly cleaned and returned to their appropriate places. All trash is to be disposed of, and tables are to be washed down with disinfectant.
13. Always wash your hands before leaving the lab.
14. Be aware of the locations of fire extinguishers, fire blankets, fire alarms, eye wash stations, and first aid kits.
15. Always wear proper attire for lab. Legs must be covered and closed shoes and socks are required. Shorts, short skirts, and sandals are prohibited. You may choose to wear a lab coat or some other kind of protective clothing. Dangling jewelry is to be removed.

MATERIALS USED IN LABORATORY EXERCISES

LABORATORY EXERCISE 1: MEASUREMENT

meter stick	weigh boats	NaCl
15 cm metric ruler	spatula	water
triple-beam balance	glass stirring rod	ice
electronic balance	thermometer	pebbles
hot plate	beakers (50 mL, 250 mL)	street map of Manhattan (see Appendix 3)
tongs	10 mL graduated cylinder	conversion tables (Appendix 10)

LABORATORY EXERCISE 2: SCIENTIFIC INVESTIGATION

triple-beam balance	beakers
thermometer	water
weigh boats	salt
stirrer	spatula

LABORATORY EXERCISE 3: SOLUTIONS

triple-beam balance	graduated cylinders (50 mL, 100 mL, 250 mL)
electronic scale	NaCl
weigh boats	glucose
spatula	wate
beakers	stock solution of 5% sodium chloride

LABORATORY EXERCISE 4: ACIDS, BASES, AND BUFFERS

Equipment:	Supplies:
pH meter	test tubes
blender	test tube rack
balance	graduated cylinders (10 mL; 50 mL; 100 mL)
Solutions and Buffers:	beakers (100 mL) pipettes
1 M HCl	pipettes
0.1 M HCl	Kimwipes

1 M NaOH   distilled water
0.1 M NaOH   pH indicator strips
1 M Tris-HCl   lemon juice
1 M phosphate buffer   vinegar
pH 4 standardization buffer   laundry bleach
pH 7 standardization buffer   dissolved baking soda
    red cabbage
    cheesecloth (sieve)

LABORATORY EXERCISE 5: MACROMOLECULES I—CARBOHYDRATES

Reagents:
Benedict's reagent
Barfoed's reagent
Lugol's iodine
Test Solutions:
1% sucrose, glucose, fructose, lactose, starch, water
Common substances:
potato juice, onion juice, Karo syrup, white bread, apple
unknown carbohydrate solution(s)
unknown carbohydrate solids

Supplies:
test tubes
hot plates
beaker tongs
test tube holders
test tube rack
marking pencil
stirring rod
spot plate
wax pencil or marker
disposable pipettes
10 ml. graduated cylinders
    600 mL beakers

LABORATORY EXERCISE 6: MACROMOLECULES II —LIPIDS AND PROTEINS

Reagents:
Sudan IV powder
bovine serum albumin (12 mg/mL)
Biuret reagent
Test Solutions:
water
oil
egg albumen
histidine
trypsin
dilute gelatin
Common substances:
white bread
unknown lipid solution 1
unknown protein solution 1
unknown protein solution 2
unknown protein solution 3
unknown protein solution 4

Supplies:
test tubes
disposable test tubes test tube racks
wax pencil or marker
brown paper towel
1 mL pipettes
5 mL pipettes
pipette dispenser
hot plates
beaker tongs
test tube holders
stirring rods
spectrophotometer
cuvettes
holder for cuvettes
parafilm
Kimwipes
vortex

LABORATORY EXERCISE 7: MICROSCOPE

Olympus CX31 compound light microscope
dissecting microscope
lens paper
microscope slides
cover slips
razor blade

forceps
pipettes
water
newspaper
hair
pond or river water
Planaria (flatworm)

LABORATORY EXERCISE 8: PROKARYOTES AND EUKARYOTES

light microscope
clean slides
cover slips
immersion oil
Bunsen burner
test tubes
test tube racks
small pipettes
micropipette and tips
wire inoculating loop
crystal violet or methylene blue stain
Congo Red dye

prepared slides of bacteria
prepared slides of amoeba and paramecia
Bacillus cultures
agar slants of *Bacillus and E. coli*
agar slants of yeast
bread mold
live specimens of protists
live nematode, *C. elegans*
Elodea leaves
Tetrahymena grown on 2% protease
peptone broth for 2 to 3 days
one package active dry yeast
methylcellulose

LABORATORY EXERCISE 9: THE CELL

photographs and tables within Laboratory Exercise 9

LABORATORY EXERCISE 10: DIFFUSION AND OSMOSIS

light microscope
dissecting microscope
slides
cover slips
balance
beakers (150 mL)
graduated cylinders (15 mL)
test tubes
test tube rack
marking pencil
pipettes
distilled water (room temp,
    ice, & hot)

Celsius thermometer
stopwatch
agar plates
forceps
meter stick
metric ruler
spoon
scalpel blade
paper towels
dialysis tubing
twine
disposable gloves
agar blocks of various sizes
    prepared with phenolphthalein

methylene blue crystals
potassium permanganate
    crystals
ammonium
40% maltose solution
0.1 M NaOH

saline (0.9% NaCl)
1.5% NaCl solution
pond water
animal blood
Elodea
flatworm

LABORATORY EXERCISE 11: ENZYMES

water bath
hot plate
spectrophotometer
cuvettes
conical flasks
beakers
graduated cylinders
pipettes
glass plate
test tubes
test tube rack
marking pencil

pipettes (1 mL, 5 mL)
pipette bulbs
Kimwipes
timer
thermometer
distilled water
dinitrosalicylic acid
Lugol's iodine
0.02 M sodium phosphate buffer (pH 6.9)
potato starch powder
fungal alpha amylase
maltose

LABORATORY EXERCISE 12: PHOTOSYNTHESIS

Whatman No. 1 chromatography filter
    paper cut into strips
spinach leaf pigments extracted in acetone
capillary tube
chromatography solvent of petroleum
    ether/acetone
large test tubes
rubber stoppers with hooks
ventilation hood
spectrophotometer
blender
graduated pipettes with rubber tubings
cellophane paper (red, yellow, green, blue,
    and violet)

cuvettes
ring stands and clamps (large and small)
light source
cheesecloth
razor blade
beakers
100 mL graduated cylinder
5% sodium bicarbonate
tape
Kimwipes
spinach leaves
Elodea leaves (that have been kept in the dark*)

LABORATORY EXERCISE 13: CELLULAR RESPIRATION AND FERMENTATION

water bath set at 42°C
thermometer
fermentation tubes
600 mL beakers
50 mL beakers pipettes
25 mL graduated cylinder
glass stirrers
distilled water
pipettes

6 different juices (including lemon, orange, grape)
active dry yeast
pH paper
ice bath
one fresh apple (Macintosh or Red Delicious)
knife
paper towel
sucrose
juices, warmed to 42°C

LABORATORY EXERCISE 14: CELLULAR DIVISION

water bath set at 60°C
light microscope
prepared slides
microscope slides
cover slips
onion with root tip
microtube
double-welled plastic trays
    or petri dishes
forceps
razor blade
metric ruler

1 mL pipette
chromosome separation kit
red beads
yellow beads
2 small pieces of yarn or string
1 larger piece of yarn or string
dissecting microscope
depression slides
sea urchin gametes
    (if available)
pipette
Feulgen stain
1 M HCl

water
45% acetic acid
video of sea urchin
    fertilization and early
    development

LABORATORY EXERCISE 15: DNA ISOLATION AND MODELING

LB/amp culture of bacteria transformed with
    pUC19 (grown overnight)
*Resuspension buffer(100 μL solution A + 10 μL
    RNase 800 mM NaOH 4% SDS
neutralization buffer (5M potassium acetate,
    pH5.5)
* Solution A: 25 mM Tris-HCl-pH8.0

microcentrifuge tubes
microcentrifuge
micropipetters (P1000, P200, P20)
micropipette tips
vortex

ice cold isopropanol
70% ethanol

LABORATORY EXERCISE 16: TRANSCRIPTION AND TRANSLATION

DNA gene sequence for beta-globin
genetic code
pipe cleaners
beads (20 different colors)

color-code for amino acids
drinking straws
scissors

LABORATORY EXERCISE 17: RESTRICTION ENZYMES AND FINGERPRINTING

colored pencils

scissors
glue
microcentrifuge tubes
electrophoresis apparatus
power supply
plasmid DNA
NE buffer

sterile water

restriction enzymes
pre-poured 1% agarose gels
micropipetters
micropipetter tips
50x TAE buffer,
molecular weight marker
10x loading dye

LABORATORY EXERCISE 18: DNA FINGERPRINTING

colored pencils
scissors
glue
DNA probe sequence at the end of this Exercise

LABORATORY EXERCISE 19: DNA CLONING AND BACTERIAL TRANSFORMATION

Appendix 9 (restriction sites)
colored pencils
scissors
glue
37°C incubator
UV penlight
water bath set at 42°C
micropipeterer (P20)
pipette tips (20 µl)
2 mL micropipeterer

competent *E. coli* HB101
pGLO plasmid
50 mM $CaCl_2$ solution (pH 6.1)
sterile LB plates (LB, LB/amp, and LB/amp/ara)
sterile LB medium
sterile loops
sterile transfer pipettes

LABORATORY EXERCISE 20: GENETICS

photograph of a chromosome spread
pencil
scissors

glue stick
graphic of a human karyotype
human karyotype form

STREET MAP OF LOWER MANHATTAN

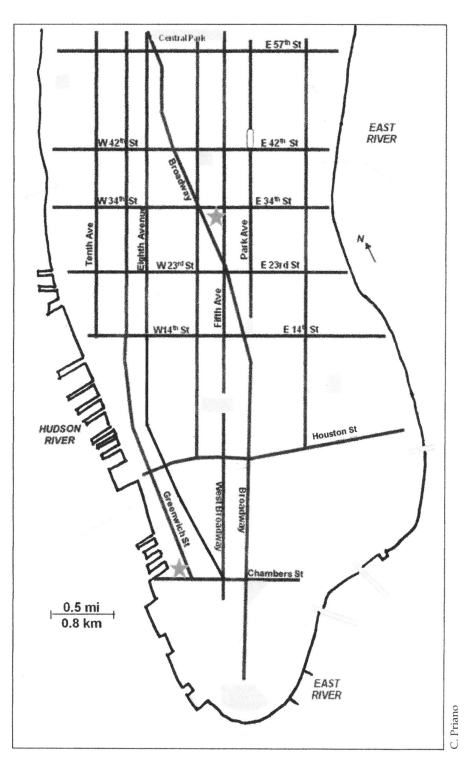

C. Priano

THE LABORATORY REPORT

Exercise 2 (Scientific Investigation) explains the process of scientific inquiry, how to design an experiment, and how to record and analyze experimental results. After a scientific experiment or set of experiments is completed, the research is documented in an organized laboratory report that is available for other scientists to review. A lab report summarizes the experimental design and explains all results, conclusions, and the significance of the project. Writing a laboratory report simulates the writing of research manuscripts for publication in scientific journals. Scientific papers should be written in the third person and should follow the American Psychological Association (APA) guidelines for formatting and referencing.

Organization is important. A properly written lab report contains the following components:

Title: The title of a lab report describes the topic of the experiment. It should clearly reflect a question that is being asked about the "observation of a phenomenon." The title can be written as a statement, or in the form of a specific question.

Using the example SO_2 experiment in Exercise 2, a title for the lab report could read "The effect of air pollution on the growth of plants." An alternative title might be "How does sulfur dioxide affect the growth of bean plants?"

Introduction: This section clearly states the purpose or aim of the experiment and why it is being done. An introduction should also give the reader relevant background information about the topic being investigated and why it is important. It should include any additional information needed to understand the experiment or experiments.

Using the same example, the purpose might read as follows: The aim of this experiment was to study the direct effect of sulfur dioxide, or SO_2, on plant growth. Sulfur dioxide is a noxious gas that is released into the atmosphere naturally by volcanic eruptions, as well as by industrial activity and the burning of fossil fuels. Not only is it harmful when breathed in, but it contributes to acid rain, which has damaging effects on plant and animal life. The experiment described in this report was designed to test whether SO_2 in the air has a direct effect on plant growth. Bean plants were selected because they grow rapidly in indoor environments and their height can be easily measured.

Hypothesis: This is a prediction about what you think the answer will be to the original question. There could be one or more possible explanations that will answer a question. In any given experiment, only one explanation or hypothesis is offered. The hypothesis is the investigator's educated guess based on common sense or on known laws and theories of science. It can be written in the form of *"The prediction was that ..."* or it might contain an *"If..., then"* statement. In the lab report, the hypothesis can be written as an isolated section, or can be included at the end of the Introduction.

For the SO₂ experiment, one hypothesis could be: "The prediction was that SO₂ would inhibit plant growth because ...", or "If a plant is exposed to polluted air, its growth is expected to be inhibited because ...". An alternatively hypothesis might be: "The presence of sulfur dioxide was expected to have no effect on the growth of bean plants because ...".

Materials and Procedures: This section summarizes the scientific experiment that tested your hypothesis. Here, you will summarize the materials and procedures so that anyone else can reproduce your experiment. If these details are already listed in your laboratory manual, then you can simply reference the laboratory manual, authors, and the pages that contain this information. However, any changes that have been made in the procedures must be noted. In this section, all variables and controls for the experiment are also identified.

In our experiment, the independent variable was SO₂ and the dependent variable was plant height. Controlled variables included the soil, water, sunlight, and temperature. In a control experiment, plants were grown in the same way using pure filtered air instead of SO₂.

Results: The results or observations of an experiment are always documented and often presented in the form of tables, charts, pictures, or diagrams. It is useful to analyze numerical data by constructing a graph. A narrative describing the results should accompany tables and charts.

Tables: A table is a useful tool for organizing data. A table can also be used to display results when there are several dependent variables (for an example of a table, see the Sample Lab Report below). When constructing a table for a report, apply the following guidelines:

1. Each table should be numbered in sequence as Table 1, Table 2, etc.
2. Each table should have a title that describes the type of results.
3. All values of the same kind should be placed within a single column.
4. A heading for each column should indicate which variable was measured and the units used for measurement.
5. The data from the experimental control should be included.

Charts and Graphs: Charts and graphs are used to show the direct relationship between the independent variable (*SO₂*) and the dependent variable (*plant height*). This not only gives a visual summary of the results, but a clear picture of the effect of the independent variable upon the experimental results. A chart can be presented in the form of a line graph, a bar graph, or a pie chart (for an example of a graph, see the Sample Lab Report below). When constructing a line or bar graph for a report, apply the following guidelines:

1. All charts and graphs, as well as other figures in a report, should be kept separate from tables, and numbered in sequence as Figure 1, Figure 2, etc.
2. The graph must contain a title that describes what is being shown.
3. The independent variable is placed on the x-axis (the abscissa or horizontal axis). If the independent variable (*SO₂*) is measured over increments of time, then plot "time" on the x-axis to represent this variable at different time increments.
4. The dependent variable is placed on the y-axis (the ordinate, or the vertical axis).

5. Both axes should intersect at an appropriate coordinate, and the numerical range for each axis should be selected so that the interpretation of the graph is clear. For a linear scale, the increments marked on the axis should be equally spaced.

6. The x- and the y-axes must be labeled, and the units of measurement need to be indicated.

7. A legend should be included to indicate the source of the data.

8. Depending on the type of data, either a line graph or a bar graph should be used: If the data are continuous, a line graph should be selected. If the data contain discontinuous measurements or non-numerical categories, a bar graph can be used. Alternatively, a pie chart might be more appropriate.

Results are valid, or believable, only if the data can be reproduced. Note that although you will only conduct a lab experiment once, the results are considered reproducible if other lab groups obtain similar outcomes. *[The experiment was repeated two more times and similar results were obtained. Therefore, the experimental results are likely to be valid].*

Discussion: The discussion is the written interpretation of the results. It attempts to explain what the charts and graphs mean. *[The poor growth of the plants exposed to SO_2 could possibly be the result of impaired cell division or rapid cell death].* Also consider whether the experiment worked. Does the data answer the original question stated in the Introduction? If the results do not answer your original question, what might be the reason? What did you learn by doing this experiment?

Conclusion: Your conclusion should state whether your results support your hypothesis. If yes, what would you do next? If no, can you think of another hypothesis to test? *[The results support the idea that air that is polluted with sulfur dioxide inhibits the growth of bean plants. Therefore, my hypothesis was supported. I would next test the effect of exposing the same plants to increasing concentrations of SO_2.]*

References: If outside references have been used to prepare any part of your report, you must include them in a bibliography according to APA formatting guidelines.

SAMPLE LABORATORY REPORT
TITLE: THE EFFECT OF AIR POLLUTION ON THE GROWTH OF PLANTS

Introduction: The aim of this experiment is to study the direct effect of sulfur dioxide, or SO_2, on plant growth. Sulfur dioxide is a noxious gas that is released into the atmosphere naturally by volcanic eruptions, as well as by industrial activity and the burning of fossil fuels. Not only is it harmful when breathed in, but it contributes to acid rain, which has damaging effects on plant and animal life ("Acid Rain", 2008). This experiment was performed to test whether SO_2 in the air has a direct effect on plant growth. Bean plants were selected because they grow rapidly in indoor environments and their height can be easily measured. Because of the damaging effect of acid rain on plants, it was predicted that SO_2 would inhibit or stunt the growth of the plants.

Materials and Procedures: The materials used for this experiment were as follows: 100 bean plant shoots of equal size planted in equal amounts of soil; two ventilation chambers (each 2 m long × 2 m wide Palatino 2 m high), one ventilated with pure filtered air and one ventilated with a fixed concentration of SO_2; tap water; and a metric ruler for measuring height of plants.

For this experiment, 50 bean plants were placed in the chamber ventilated with SO_2. As a control, 50 bean plants were placed in the second chamber ventilated with pure filtered air. Both chambers were situated where they would receive the same amount sunlight. Each set of plants was watered once every day with an equal amount of water. On day 0, the height of each plant shoot was measured in cm from the soil level to the top of the shoot. For each plant, measurements were recorded every five days for one month. The average height was calculated for plants in each of the two chambers and analyzed as a function of time exposed to either SO_2 or pure air.

The independent variable for this experiment was SO_2 and the dependent variable was plant height. All other variables were controlled, including the type and amount of soil, water, sunlight, temperature, and original shoot size.

Results: The results of this experiment are summarized in Table 1 and Figure 1. In the presence of SO_2, plant growth was inhibited by day 5 whereas in the absence of SO_2, growth steadily increased. After one month, the plants exposed to SO_2 grew to an average of only 8.7 cm, whereas the control plants grew to an average of 30.8 cm. The experiment was repeated two more times and similar results were obtained, therefore the results are believed to be valid.

Discussion: The poor growth of the plants exposed to SO_2 could possibly be the result of impaired cell division while plant cells are multiplying. Alternatively, SO_2 might interfere with other essential physiological processes, such as photosynthesis. The results suggest that an increase in sulfur dioxide levels in our atmosphere from the burning of fossil fuels will cause increasing damage to plant life in our ecosystems.

TABLE 1 Average height of bean plants.

Air ⟶ / ↓ Day	Shoot length (cm) using clean air (control)	Shoot length (cm) using SO$_2$ polluted air
0	5	5
5	6.2	5.6
10	8.8	6.8
15	11.5	6.7
20	17.1	7.3
25	21.7	7.8
30	30.8	8.7

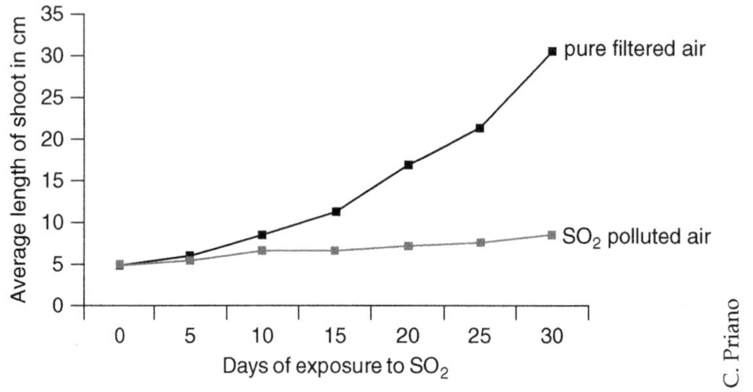

FIGURE 1. Effect of SO$_2$ on the average height of bean plants.

Conclusion: In support of the initial hypothesis, air that was polluted with sulfur dioxide inhibited growth of the bean plants. Future experiments will be designed to test the effect of exposing the same type of plants to increasing concentrations of SO$_2$.

Reference: "Acid Rain." (2008). U.S. Environmental Protection Agency, 2008. Retrieved from http://www.epa.gov/acidrain.

APPENDIX 5

THE pH METER

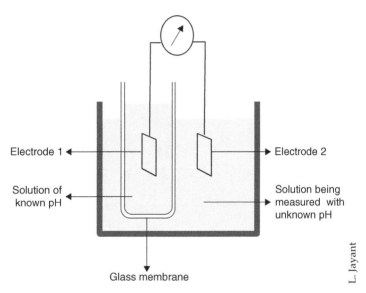

FIGURE A5.1   Principle of the glass electrode method.

A pH meter (Figure A5.1) is a device that is used to measure acidity or alkalinity of a solution accurately. A pH meter consists of a pH probe and a voltmeter. The pH probe senses changes in the hydrogen ion concentration in a solution. The probe consists of an internal electrode or glass electrode and a reference electrode. The reference and glass electrodes are both wires coated with silver chloride and each are kept in a separate compartment called a half cell. While the reference electrode is immersed in saturated KCl solution, the glass electrode is immersed in 0.1M HCl. The tip of the probe is a glass membrane that is sensitive to changes in hydrogen ion concentration. The voltmeter is a high resistance device that precisely measures the potential difference between the reference electrode and the glass electrode.

The pH probe measures the electrochemical difference between the liquid inside the glass electrode and the unknown liquid outside. Since the thin glass bulb allows the hydrogen ions to react with it, the glass electrode essentially measures the electrochemical potential of hydrogen ions in the solution outside. The reference electrode is not sensitive to changes in hydrogen ion concentration and serves as a reference. The potential difference between the glass and the reference electrodes measures the hydrogen ion concentration or pH of the solution.

USE OF THE SPECTROPHOTOMETER

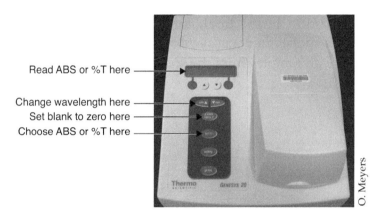

Read ABS or %T here

Change wavelength here
Set blank to zero here
Choose ABS or %T here

O. Meyers

FIGURE A6.1.   Genesys 20 Spectrophotometer.

A spectrophotometer (Figure A6.1) measures the ability of a compound to absorb or transmit selected wavelengths of light. The ability of a compound in solution to absorb light (absorbance) is proportional to the amount of compound present and therefore enables a means of determining concentration. The following instructions are specific for measuring absorbance (ABS) of compounds in solution. A blank is always needed to standardize the spectrophotometer. The blank is a control sample that contains everything except the compound that is being measured. Below are both instructions for using a Genesys 20 spectrophotometer (Figure A6.1) and general instructions for measuring absorbance (ABS) on any visible light spectrophotometer.

Instructions for using the Genesys 20 Spectrophotometer

1. Turn power button (rear of instrument) to "ON".
2. Press A/T/C button to either select Absorbance or % Transmittance or Concentration mode.
3. Press nanometer, either nm↑ or nm↓, to select wavelength.
4. Open lid and insert cuvette with blank, then close lid.
5. Press 0 ABS/100% T button. This will set the blank at 0 absorption or 100% transmittance depending on the scale you chose in step 2.
6. Remove blank and insert cuvette with your sample. Close lid and read and record appropriate value on display.

General Instructions for using a spectrophotometer to measure absorbance (ABS)

1. Turn the power button to "ON". Allow the spectrophotometer to warm up for 15 minutes.
2. Set the Mode button to "Absorbance".
3. Set the appropriate wavelength in nanometers (nm) according to your Procedure.
4. Open the lid of the cuvette chamber and insert a cuvette containing your blank. Close the lid.
5. To standardize the spectrophotometer, set the readout for the blank at 0 Absorbance (ABS)/100% Transmittance (T). The instrument will be standardized for the wavelength set in step 3.
6. To obtain the ABS value of your test sample, remove the blank and insert a cuvette containing your test sample. Close the lid and record the ABS value shown on the display. This is the absorbance value of your test sample at the set wavelength.
7. If subsequent samples are to be read at the same wavelength, insert each cuvette one at a time and record the ABS reading for each.
8. If samples are to be read at different wavelengths, you must first restandardize the spectrophotometer for each wavelength. Repeat step 3 to change the wavelength; using the same blank, repeat step 4 to restandardize the spectrophotometer for the new wavelength; and repeat step 6 to read the absorbance (ABS) of the experimental sample at the new wavelength.
9. When finished, remove all samples from the spectrophotometer and turn the power button to "OFF".

APPENDIX 7

ELECTROPHORESIS

The process of separating charged molecules using an electric current is called electrophoresis. All DNA molecules carry a negative charge because of the negatively charged phosphate backbone. They will move toward the positive electrode (anode) in an electric field. Therefore, charge alone cannot be used to separate DNA fragments.

Gel electrophoresis is a procedure used for separating DNA molecules through a stationary gel-like material when an electrical field is applied. This procedure separates DNA molecules based on their rate of movement through the gel. There are several factors that influence the rate of movement of DNA molecules in an electric field: (1) strength of the electrical current, (2) the concentration of the gel matrix, and most importantly, (3) the size of the DNA molecules. In general, smaller DNA molecules move through the meshy pores of a gel faster than larger molecules.

Gels are made of materials such as polyacrylamide or agarose. Agarose is a polysaccharide derived from seaweed. Agarose gels are used to separate DNA fragments larger than 500 base pairs long, whereas polyacrylamide gels are used to separate smaller DNA fragments.

Agarose is insoluble in water. However, at high temperatures, it forms a colloid with water. When cooled, the colloid takes the shape of the mold it is stored in (like Jell-o). The mold, or gel tray is equipped with a comb that will make wells in the gel. These wells serve as reservoirs to hold the DNA sample prior to the electrophoresis. The gel is placed in a buffer, which serves to conduct an electric current from one end of the gel to the other. As the current runs through the gel, DNA, which is negatively charged, moves toward the positive electrode. As the DNA moves, different-sized pieces or fragments will separate according to size.

DNA itself is not visible within an agarose gel. After electrophoresis, DNA is visualized by the use of a dye that binds to the DNA fragments in the gel. Some dyes, such as ethidium bromide and Syber green, fluoresce under ultraviolet (UV) light and can be used to visualize very small amount of DNA easily, but these dyes can be very toxic. Other dyes are safer to use, such a methylene blue, but are not as sensitive.

Preparing 1% agarose

1. Weigh out 1 g of agarose.
2. Suspend in 100 mL TAE buffer* (see below).
3. Heat the mixture in a microwave until it begins to boil.
4. Cool the mixture to 45°C.
5. Pour into a prepared gel tray with the comb in place.

247

Pouring the agarose

1. Prepare the tray in which the agarose colloid will be poured. The edges of the tray must be sealed with tape or have a ledge that is leak-proof. See Figure A7.1 below.
2. Place the comb in the grooves on the sides of the tray.
3. Pour the cooled agarose (45°C or lower) into the tray.
4. Allow the gel to solidify.
5. Gently remove the comb by pulling straight up. Do not remove it forcefully. This will damage the wells.

Setting up the electrophoresis apparatus and running a gel

1. Place the gel in an electrophoresis tank as shown in Figure A7.2.
2. Fill tank with TAE electrode buffer (see below). The purpose of this buffer is to conduct electricity from on electrode to another and complete an electrical circuit.
3. Add enough buffer solution to submerge the gel completely and fill the wells.
4. Add 5 µL of loading buffer** (see below) to your DNA sample and mix. Loading buffer contains glycerol, making it more viscous than the electrophoresis buffer, and bromophenol blue, a blue dye that can be used to track the sample during electrophoresis.
5. Carefully pipette your samples into the wells of the gel. The viscous loading buffer will allow your sample to fall to the bottom of the well.
6. After loading the wells, place the lid on the gel box, connecting the electrodes on the box.
7. Connect the electrode wires to the power supply, making sure the positive (red) and negative (black) electrodes are connected correctly. Remember, DNA will migrate toward the positive electrode.
8. Turn on the power supply and set the voltage to 100 V. Make sure the current is running through the buffer by looking for bubbles forming on each electrode.
9. Check the gel to make sure that the DNA is moving in the correct direction by observing the movement of the blue loading dye. If the dye is moving in the wrong direction, turn off the power supply, unplug the electrodes, and fix your setup (you might have to reverse the electrodes). Make sure the gel does not get hot at any time.
10. Run the gel until the blue loading dye has moved three-fourths the length of the gel.
11. Turn the power off, unplug the electrodes, and gently remove the lid. Carefully lift the gel from its tray using gloved hands.

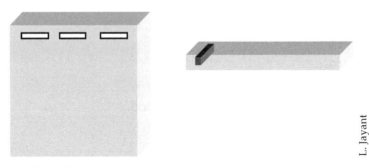

L. Jayant

FIGURE A7.1   Agarose gel.

12. Put the gel in a tray containing the stain of choice, preferably DNA "Fast Blue" stain (methylene blue).

13. Stain the gel 15 min.

14. Pour the dye back into its container (the dye can be reused many times).

15. Add cold water to the gel and let it destain for 15 minutes. The stain will wash out of the gel, except where it is bound to DNA.

16. DNA fragments will look like blue bands against a white background when viewed on a white light box or on a white piece of paper.

Electrophoresis Buffers:
*TAE buffer (Tris-Acetate/EDTA):

1. Add 4.84 g of Tris base to ~900 ml H2O.

2. Add 1.14 ml glacial acetic acid and 0.5 M EDTA (pH 8.0) to the solution and mix.

3. Pour mixture into a 1 L graduated cylinder and add H2O to a total volume of 1 L.

4. Note—for convenience a concentrated stock of TAE buffer (either 10X or 50X) is often made ahead of time and diluted with water to a 1X concentration prior to use.

**6X sample loading buffer

1. 1 mL sterile H2O

2. 1 mL glycerol

3. 0.05 mg bromophenol blue

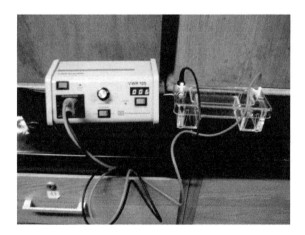

FIGURE A7.2   The electrophoresis apparatus.

SOLUTIONS TO COMPREHENSION CHECKS

LABORATORY EXERCISE 1: MEASUREMENT

Part I:
1) 70
2) 900
3) 8000
4) 5000
5) 5000
6) 2000
7) 40,000
8) 3000
9) 60,000
10) 2800
11) 49,000
12) 16,000
13) 5000
14) 140
15) 4200
16) 35,000
17) 34,000,000
18) taller; 5
19) 1000
20) Leona; 100 m
21) 3000; Susan; 2600 m
22) Charles; Lynsee
23) 10
24) 10,000
25) penny; 1 g

Part II:
1) 48
2) 8
3) 15
4) 5280
5) 54
6) 8
7) 46
8) 100
9) 118
10) 71
11) 108
12) 12
13) 360
14) 22
15) 14
16) 72, 70; Becky; 2 ft

Part III:
1) 3
2) 2
3) 2
4) 7
5) 18
6) 102
7) 14
8) 9
9) 3
10) 22
11) 15
12) 11
13) 12; 0.75; 2.84
14) 3, 4; Sara; 1

LABORATORY EXERCISE 3: SOLUTIONS

1) 20 %

2) 0.25%

3) 25 g

4) 0.1875 g or 187.5 mg

5) 500 mL

6) 48 mL

7) 5 M

8) 2.32 moles

9) 9.855 grams

10) 2.5 M

11) 0.5 M

12) 0.2 M

13) 4 liters; 6 liters

14) 1 mL; 0.2 mL

15) 10%; 1 mL

16) 0.2 g

17) 584.4 g

18) (a) 10 mL (b) 10 mL (c) 25 mL

19) 10 mL

20) (a) 6 g Tris-HCl (b) 2.9 g NaCl (c) 2.5 g SDS (d) 93 g EDTA

APPENDIX 9

RESTRICTION ENDONUCLEASES

ENZYME	RECOGNITION SITE
Alu l	5′AG▼CT 3′
BamH l	5′G▼GATCC 3′
Bg/ Il	5′A▼GATCT 3′
BssH II	5′G▼CGCGC3′
EcoR I	5′G▼AATTC 3′
EcoR V	5′GAT▼ATC 3′
HaeIll	5′GG▼CC3′
Hha I	5′GCG▼C 3′
Hind III	5 ′A▼AGCTT 3′
Hpa I	5′GTT▼AAC 3′
Hpa II	5′C▼CGG 3′
Kpn I	5′GGTAC▼C 3′
Nco I	5′C▼CATGG 3′
Nhe I	5′G▼CTAGC 3′
Pst l	5′CTGCA▼G 3′
Sal l	5′G▼TCGAC 3′
Sca l	5′AGT▼ACT 3′
Sma l	5′CCC▼GGG 3′
Xho I	5′C▼TCGAG 3′
Xma I	C▼CCGGG

APPENDIX 10

CONVERSION TABLES

TABLE A10.1 Unit conversions within the U.S. system of measurements.

Length/Area	Weight	Volume
1 ft = 12 in	1lb = 16 oz	1 tbsp = 3 teaspoon (tsp)
1 yd = 3 ft	1 ton = 2000 lb	1 c = 16 tbsp
1 mi = 1760 yd		1 pt = 2 c
1 sq ft = 144 sq in		1 qt = 2 pt
1 sq yd = 9 sq ft		1 gal = 4 qt
1 acre = 4840 sq yd		
1 sq mi = 640 acres		

TABLE A10.2 Unit conversions between the U.S. and SI systems of measurement

Length/Area	Weight (mass)*	Volume
1 inch = 2.54 cm	1 oz = 28.35 g	1 fluid oz = 30 mL
1 ft = 0.3048 m	1 lb = 0.45 kg	1 qt = 0.95 L
1 mi = 1.609 km		1 gal = 3.785 L
1 sq ft = 0.0929 m^2		1 sq mi = 2.59 km^2
1 cm = 0.39 in	1 g = 0.035 oz	1 L = 1.06 qt
1 mi = 1.61 km	l kg = 2.2 lb	
1 km = 0.62 mi		

*weight = mass x gravity. Since the gravity of the earth is constant, the mass of an object in grams is proportional to its weight in pounds.